D1443364

Private Branch Exchange Systems and Applications

Other McGraw-Hill Communications Books of Interest

To order or receive additional information on these or any other McGraw-Hill titles, in the United States please call 1-800-822-8158. In other countries, contact your local McGraw-Hill representative. **MH93**

Private Branch Exchange Systems and Applications

Stanley E. Bush

Charles R. Parsons

McGraw-Hill, Inc.

New York   San Francisco   Washington, D.C.   Auckland   Bogotá
Caracas   Lisbon   London   Madrid   Mexico City   Milan
Montreal   New Delhi   San Juan   Singapore
Sydney   Tokyo   Toronto

Library of Congress Cataloging-in-Publication Data

Bush, Stanley E.
  Private branch exchange systems and applications / Stanley E.
Bush, Charles R. Parsons.
      p.   cm.—(McGraw-Hill series on computer communications)
  Includes index.
  ISBN 0-07-009706-2
  1. Telephone—Private branch exchanges. I. Parsons, Charles R.
II. Title. III. Series.
  TK6397.B88   1993
  621.385—dc20   93-2138
                                                              CIP

1 2 3 4 5 6 7 8 9 0   DOC/DOC   9 9 8 7 6 5 4 3

ISBN 0-07-009706-2

*The sponsoring editor for this book was Stephen S. Chapman and the
production supervisor was Donald F. Schmidt. It was set in Century
Schoolbook by North Market Street Graphics.*

Printed and bound by R. R. Donnelley & Sons Company.

Dimension® is a trademark of American Telephone and Telegraph Com-
pany.
Dimension System 85® is a trademark of American Telephone and Tele-
graph Company.
Touchtone® is a trademark of American Telephone and Telegraph Com-
pany.
UNIX® is a trademark of American Telephone and Telegraph Company.
DOS® is a trademark of Microsoft Corporation.
Meridian® is a trademark of Northern Telecom Inc.
Definity® is a trademark of American Telephone and Telegraph Company.

Contents

Preface

During the past 25 years, while the authors designed PBX systems and applications to run on them, very few books have been devoted to the explanation of the structure and the use of Private Branch Exchanges. Since PBX systems are a key element in the voice communication network of many companies of medium to large size, and since they provide access to the public network for many important data connections, they are a key asset that may be deployed and managed to facilitate the strategic plans of the enterprise. It is therefore surprising that they have received so little attention in print. Engineers, telecommunication managers, students, and sales personnel have had to rely on information from vendors, brief descriptions in books on other telecommunication subjects, or occasional articles in the trade magazines.

This book is aimed at fulfilling the needs of a broad cross section of telecommunications professionals who require more information about PBX system architecture and applications. It provides information about the technology, architecture, features, operation, and management of a modern PBX system. The material is presented in a practical way so that it may be directly applied in selecting and managing a PBX system or in building a network of PBX systems.

The original purpose of PBX systems was to reduce the number of wires necessary to provide telephone service to business enterprises. PBXs followed the development of the central offices available at the time with the only distinguishing characteristic being that of smaller size. Over the years PBXs have evolved into independent systems that are used to control private network voice and data communications and to access the public network. Modern PBX systems provide virtually hundreds of features to manage and control communications. The complexity of modern PBXs along with the divestiture of the Bell System have brought about a situation where the selection, operation, and maintenance of a PBX system is a reasonably complicated task.

This book is intended to help the telecommunications professional understand the operating characteristics and capabilities of today's

PBXs. You should find it useful in determining how best to manage your communication system with minimum risk while providing the required capability and flexibility at a reasonable cost. The book sets the stage for a discussion of modern PBXs by briefly discussing the history of PBX systems. It also provides tutorial material on the structure and operations of a PBX system. The first half of the book discusses the history, hardware, and architecture aspects of PBXs. The second half then discusses the features, operations, maintenance, and security issues associated with PBX systems. In addition, the book also constitutes a valuable reference source for evaluating PBX purchases and comparing vendor alternatives. It will help you match available PBX systems to your specified current needs, and it will help you select a system that will be flexible enough to remain viable over the time frame of your planning horizon.

This book may be used by anyone interested in the subject as a general introduction to the world of PBX systems. It can also be used as a general reference to answer questions about the architectural or feature differences among PBXs and the importance of such differences. It will be found useful by many in preparing Requests For Proposal. While this book is not intended as a textbook, it may be used in teaching telecommunications as a supplement to other texts that generally lack significant material on PBX systems and their uses.

Significant changes are about to take place in the PBX field. Systems will become more and more distributed and will incorporate new standard technologies such as ISDN, Frame Relay, and Asynchronous Transfer Mode. It is the authors' sincere hope that this book will help you track these trends so that you will be able to avoid being stuck with a "white elephant" or straying over the "bleeding edge" of PBX technology.

Acknowledgment

The authors would like to thank Dan Callahan who originally pointed out the need for a book on PBX architecture and applications. Dan started writing a few chapters before his booming consulting business swept him away from the project. We hope that this book at least resembles what he had in mind.

Stanley E. Bush
Charles R. Parsons

Market Positioning

Introduction

While the telephone itself is now over 100 years old, the *PBX*, or *Private Branch Exchange,* came along much later. It grew out of the need to provide a switching system for a single customer. However, the first PBXs were simply smaller copies of the larger central office switches. Only after the arrival of automatic switching in the late 1920s and early 1930s was the PBX recognized as a separate product.

A Private Branch Exchange is a switching system. The word exchange is simply an early term for a system that connects telephones to other telephones or to transmission facilities going to other switches. The second characteristic of a PBX is that it is private. Originally, this meant that the switch provided service for only one customer. It did not mean that the customer owned the switch. Until certain court decisions made in the late 1960s and early 1970s, the Bell System only leased equipment to customers. The word branch in PBX meant that the switch was separate and perhaps remote from the normal central office switches. Thus, a PBX is a small switching system, serving a single customer, which is distinct and possibly remote from the public switches in the network.

As a result of deregulation, customers may either purchase or lease a PBX. In addition, there are arrangements wherein one customer may own a PBX and lease some of the capacity to another customer. Also manufacturers, in order to distinguish their products from others, have added several letters to the acronym. PABX indicates that the PBX is automatic or controlled by dial pulses or push button signals rather than a human operator. EPABX designates an electronic version of an automatic private branch exchange. The acronym CBX stands for a computerized branch exchange. Each of these designations refers to some variation of the private switching systems known as PBXs. In this book, we will refer to PBX systems without added initials.

Very early in the history of the telephone, users discovered that it was inconvenient to have a different telephone connected to each other user. Before switching was invented, a manager may have had a telephone to call home and another to call the factory. Telephone switching systems have significantly reduced the cost and improved the convenience of connecting telephone calls. The first switch served only eight lines. It consisted of several manual switches which were used to connect one line to another. In addition, a telephone circuit allowed the operator of the switch to talk to the users who were referred to as subscribers. Because the switches were mounted on a wooden board, these primitive switching systems became known as switchboards. As the need for larger switching systems emerged, the switchboard evolved into a system of plugs and jacks. A person, called an operator or attendant, provided the intelligence to operate these cord switchboards, or *cord boards*. When a subscriber went off-hook requesting service, a small metal flag dropped next to the appropriate jack. On later cord boards a lighted lamp indicated the request for service. Usually some sound to alert the operator accompanied the visual indication. The operator inserted a plug (Fig. 1.1) attached to one end of a cord into the jack belonging to the subscriber requesting service. This established a connection to the operator's telephone circuit. The operator asked for the number and then plugged the other end of the cord into the appropriate jack to complete the connection. The operator alerted the called party by sending a ringing signal to their telephone. When the called party answered, the operator threw a switch that established the talking path. Visual and audible indications also indicated the completion of a call. The operator then removed the unneeded cords.

The earliest PBX systems were simply smaller cord boards. They worked in exactly the same way as the central office cord boards. If the

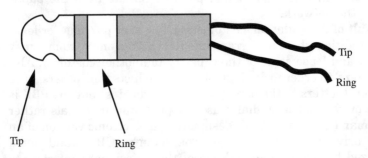

Tip

Ring

Tip   Ring

Figure 1.1   The components of a telephone plug.

subscriber wanted to customize the operation, he or she had only to tell the operator what was wanted. For example, if the customer wanted the operator to take messages for people who were busy on their line or away from their telephone, he or she simply told the operator to do so. Customizing features on modern PBX systems is considerably more difficult. These manual cord boards served well for many years. In fact, there were several in use even as late as the 1960s. It is likely that some small motel somewhere is using one today.

In the 1890s, Almon B. Strowger designed, patented, and constructed the first automatic switching system. Whether entirely true or not, the story is an interesting one. Strowger was an undertaker. The operator of the manual switching system in town was a relative of a competitor. Strowger suspected that she diverted calls to the relative unfairly. Whatever the reason, Strowger receives credit for the design that evolved into the step-by-step (SxS) switch. The basic switching mechanism has ten rows of ten contacts each. When wired together these switches form much larger switching systems. The first switch in the connection is called a line finder. When a telephone user goes off-hook, or lifts the receiver, the current flowing in the line and through the telephone creates an indication that service is required. This causes the line finder (Fig. 1.2) to step up and then across to make a connection to the line requesting service. At this point, dial tone is provided through the line finder to the telephone requesting service. When the calling party dials the first digit, the next switch in the connection, called a *selector,* steps up one row for each pulse in the digit. Thus, dialing a five causes the selector to step up five rows. The selector then hunts across the row looking for an idle switch to complete the next part of the connection. It skips any that are busy and stops at the first one that is idle. This next switch in the chain is also called a selector. If all selectors are busy, the calling party receives busy tone. If an idle selector is found, it handles the next digit in a manner similar to the first selector. That is, it steps up a number of rows corresponding to the number of pulses in the digit dialed. Then it hunts across the row to find an idle switch to complete the next part of the connection. Finally,

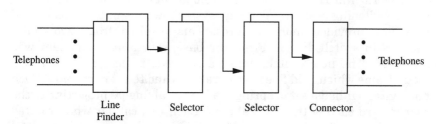

Figure 1.2 A step-by-step switching arrangement.

when there are only two digits left to dial, a connection has been established through a series of switches to the final switch called a *connector*. The connector steps up one row for each pulse in the next to last digit dialed and then waits for the final digit. The final digit causes the connector to step over one place for each pulse. The user of the subscriber line which is terminated on the connector at the point selected corresponds to the number dialed. Relays then apply ringing to the called line; when the called user answers, the connection is completed.

Step-by-step switching systems worked well and were inexpensive even though they did require some maintenance to keep everything properly adjusted. They brought with them some new terminology as well. One new term resulted from the method of making connections. Conversations are separated on different wire paths through the system. Switches which separate the connections physically are called *space-division switches*. The control is progressive, meaning that each switch handles only one digit and then passes the connection on to the next switch in line. This type of distributed control makes it impossible to select an alternate path through the system when all the paths to the next switching stage are busy. Idle paths may exist but cannot be used because there is no way to back up. Other limitations were not apparent for several years until new features such as call forwarding came along. It should be clear that a step-by-step switching system cannot route a call to some line other than the one corresponding to the number dialed. Even with these limitations, step-by-step switching systems served well for many years and many are still in place.

The Bell System offered a line of step-by-step PBX systems called the 701 series. These were the first private branch exchanges to provide automatic or dial service. They operated exactly as did their central office counterparts, only they were usually smaller. Many are still in use today. In fact, a uniquely designed 701-type PBX provided service to the White House for many years and was finally retired from service in 1987.

Common Control Systems

After World War II, research and development efforts modernized telephone switching systems. The new systems were smaller, less expensive, and provided more features. This resulted from two major changes in switch design. First, the basic switching component was improved. The new switch, called a *crossbar switch,* consisted of a metal frame which held five horizontal bars and ten vertical bars. Each horizontal bar had a set of spring-loaded metal fingers projecting to the rear toward each of the vertical bars. Each vertical bar when rotated operated a set of contacts located just above and below each horizontal bar. Thus there were ten sets of contacts on each vertical bar, and since there were also ten vertical bars, the entire switch housed 100 sets of contacts. However, when the vertical bar rotated, the contacts closed only if one of the metal fingers projecting from the horizontal bar was between the contacts and the rotating vertical bar. So a particular set of contacts closures could be initiated by first swinging a horizontal bar up or down. This put any of the spring-loaded fingers that were not in use between the vertical bars and the corresponding set of contacts. Then, rotating one of the vertical bars closed the appropriate contacts at the point where the fingers were in position. This connection stayed in place as long as the vertical bar remained in the rotated position, because the springs fingers were clamped between the vertical bar and the contacts. Each of the fingers not clamped to contacts by a vertical bar returned to the middle or unoperated position. At this point the horizontal bar was free to make other connections as required at other locations. The switch therefore had 10 independent input lines and 10 output lines. It provided the capability of connecting any input to any output for up to 10 connections at a time. Wiring these switches together in arrays formed a *switching matrix*. This switching mechanism is a space-division switch where each connection is set up on a separate metallic path.

The second major change was in the control mechanism used to operate the switching matrix. In the older manual systems the control, the operator, was completely independent from the switching mechanism, the cords. If the operator or attendant found a line to be busy they could establish a connection to some other line. This operation is a form of the modern feature call forwarding. Such an operation, as pointed out above, was not possible in a progressive control step-by-step switch. In the crossbar systems the control was separate from the switching matrix. This method was called *common control*. That is, the control system was common to the entire switching system and did not pertain to one switch only. The control element, called a *marker*, consisted of a set of relays wired in such a way that it preformed the necessary functions of establishing the proper connections (Fig. 1.3). The name marker comes from the method used to establish connections. The two ends of the connection were first electrically marked and then a connect command was issued. A set of registers assisted the marker by collecting the digits dialed. A translator converted the dialed digits to equipment numbers as required by the marker.

When a telephone requested service by going off-hook, an originating marker established a connection though the switching matrix to a *register*. The register provided dial tone and prepared to collect digits. The subscriber then dialed the desired number. When the register had collected a complete number it signaled a completing marker that it was

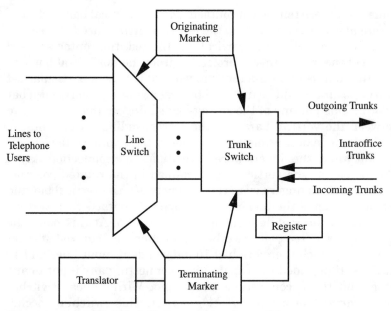

Figure 1.3 A simplified block diagram of a crossbar switch.

ready. The completing marker then made the final connection from the telephone connected to the register to the telephone represented by the number dialed and stored in the register. To do this, the completing marker enlisted the aid of the *translator*. The translator was capable of taking the actual dialed number as input and translating it into the equipment numbers needed for the connection. Thus, telephone numbers were no longer tied to the actual equipment which provided the connection. Changes were made by simply changing the information stored in the translator. There were many kinds of translators, but one frequently used consisted of a set of cards with holes punched in them. Some of the cards dropped depending on the actual number dialed. Light shining on the cards appeared at the output side of the translator where the holes line up. Changing the contents of the translator required punching new holes in a set of cards. This translation function and control independent of the switching matrix provided considerable flexibility in the way the system operated.

In the Bell System the 750 series of PBXs grew out of this technology. The common control of the 750 series systems allowed enhanced features such as dial access to conference circuits, digit translation to permit connection to a variety of special service circuits, and switched loops, more flexible connections, to the attendants.

The next phase in the evolution of switching systems came not as a result of a new architecture but as a function of technology. In the

1960s the transistor and evolving semiconductor technology made it possible to replace the relay marker with an electronic common control. This provided advantages in speed of operation, reduced initial cost, and lowered maintenance costs. In addition, where changing the program or the way in which the control operated required rewiring relays in the markers, the rewiring was much simpler in the electronic controls. In fact, features were changed by simply inserting diodes at appropriate spots in a suitably labeled matrix.

While this was going on, the switching matrix was also changing due to technological improvements. *Reed relays,* small metallic contacts in glass bottles, or semiconductor electronics replaced the crossbar switches. These networks provided the same matrix architecture as the crossbar switches did but they improved the speed of operation, lowered the initial and life cycle costs, and reduced the size and power consumption of the system.

These enhancements first appeared, as usual, in central office switching systems and later in PBX systems. In the 1960s the Bell System incorporated these technologies into the 800 series of PBX systems. These systems were smaller and more reliable than previous systems. It was easier to change options and features. In addition, it was possible to take advantage of the reductions in cost and size brought about in the semiconductor industry, at least for the system control circuits. On the other hand, reed relays wired together provided the switching network function. The circuits necessary to control the lines and trunks contained several relays and individual semiconductor parts. As a result, achieving maximum advantage from the semiconductor revolution required additional enhancements to PBX systems.

These changes also had their roots in the early 1960s with the development of switches based on computer control. In the Bell System the #1 ESS was the first electronic switching system or central office based on stored program control. A computer controlled the entire system. A memory system provided storage for the program which specified how the system should operate. Now, adding new features or changing old ones required only software changes. Also, changing options such as numbering, routing patterns, or restrictions was simply a matter of changing the translation software stored in the memory. Using a computer to control a switch was a major advance which allowed telephone switches to take advantage of many developments in the computer industry.

The other change required to take advantage of semiconductors in switching systems was the development of *time-division switching.* All previous systems up through the #1 ESS had space-division switching networks. That is, each telephone call existed on a separate physical path through the system. Some set of metallic contacts provided the necessary electrical connections. No other conversation could share any of the switching equipment used for another call.

Time-division switching, a new technique perfected in the early 1970s, allows several calls to time-share expensive switching circuits. The theory necessary for digital time-division switching was established several years earlier. However, the transistor and semiconductor electronics were necessary to make the system practical. To convert a voice signal to a digital format (Fig. 1.4) the coder/decoder (CODEC)

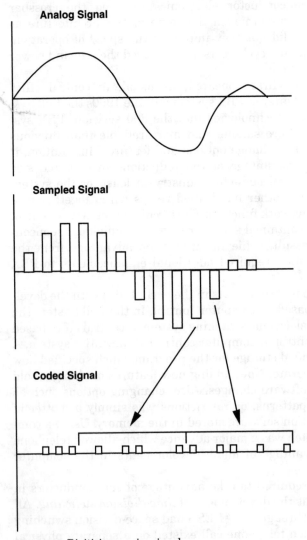

Figure 1.4   Digitizing a voice signal.

samples the analog voice input 8000 times per second. It then assigns a number to each sample based on its amplitude. The CODEC then creates a string of eight binary digits to represent the assigned number. These binary digits, or bits, may be switched or transmitted in place of the original analog signal. This digital signal is not affected by most of the noise and distortion in transmission or switching systems. Since eight bits represent each sample, and samples occur at a rate of 8000 times per second, the overall transmission rate of a standard digital voice signal is 64,000 bits per second.

Each bit of the digital signal may be represented by a very narrow pulse. As a result, it is possible to interleave the pulses from several digital signals on one transmission line. The signals occupy repetitive time positions called *time slots*. Each time slot occurs at a rate of 8000 times per second. A group of time slots is called a *frame*. In the T1 transmission system there are 24 time slots in each frame. One additional bit is added to the beginning of each frame to identify the beginning of the frame. This transmission technique is known as Time Division Multiplexing (TDM). It is also possible to take advantage of this time-division technique in performing telephone switching functions.

Switching takes place in a time-division system by interchanging the information in two of the time slots (Fig. 1.5). The information in each time slot is stored sequentially in a set of memory locations. The stored information is then read out in an order specified by a control memory. If the control is set to read out the information from telephone *A* in the time slot going to telephone *B* and *B*'s information is read into the time slot going to *A* a two-way connection is established.

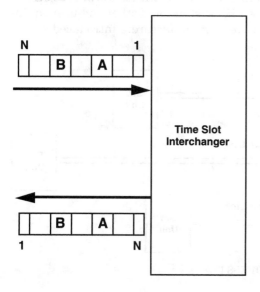

Figure 1.5 A time slot interchanger.

However, in this case the switching is done entirely in memory. It is controlled by an additional set of memory locations loaded by the processor. Each of these units is shared among all the users in the multiplexed digital stream. This combination of components is called a *time slot interchanger.*

With stored program control and electronic switching such as time slot interchanging, central offices and PBX systems could now take advantage of the cost and size reduction available in semiconductor electronics. They could also benefit from the advances in computer hardware and software. Now, PBX systems were at last designed as separate systems and not as smaller versions of a central office. There were differences in perceived needs of the PBX customers. In addition, features and services provided by the switch were simply different sets of software instructions. Thus, PBX systems and central offices could easily have quite different features and services to accommodate the different needs.

Developed in the mid-1960s, the first such PBX was the 101 Electronic Switching System (ESS). It was unique in that a major component (the control unit) resided in the local central office (Fig. 1.6). The control unit contained the processor which ran the software and housed the system administration and maintenance facilities. The switch units, located on the business premises, managed the interface with the terminals (telephone sets and peripheral devices) and the attendant (operator) consoles. Communication between the two components took place over a high-speed data link. Each control unit could service several switch units of different sizes.

The Carterfone decision in 1969 allowed other vendors to attach their customer premises equipment to the public network. This stimulated interest in designing new PBX systems and other end-user equipment. So, in the early 1970s several manufacturers introduced stored

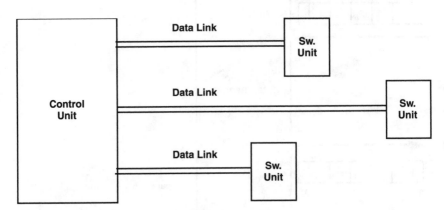

Figure 1.6   Block diagram of the #101 ESS PBX.

program control systems with time-division switching. Many of these had no relation to a former central office design. In 1975, Northern Telecom introduced the SL-1 series of PBX systems. These provided stored program control and digital time-division switching. ROLM introduced the CBX line of PBX systems which also incorporated stored program control and time-division switching. However, while the Northern products used the standard Pulse Code Modulation (PCM) version of digitized voice, the ROLM products used a nonstandard system. They sampled the voice at 12,000 times per second and used 12 bits to encode each sample. The audible performance of the system was fine but this nonstandard scheme later caused trouble in interfacing with standard PCM transmission facilities by requiring significantly more circuitry to make the connection.

At about the same time, AT&T introduced the Dimension line of PBX systems. These systems also used stored program control but they relied on an analog time-division switching technique known as Pulse Amplitude Modulation (PAM). PAM systems start by sampling the voice signal, the same as in PCM. However, they do not convert the pulses into a digital binary format. Rather, a PAM system switches the pulses directly. This system worked well for voice transmission but presented some difficulties in accommodating data at high rates.

The stored program control in these new systems allowed a variety of sophisticated features. Advanced networking features allowed multilocation businesses to integrate their communications by using unified numbering schemes. Calls terminating at busy or unanswered telephones could be redirected to other locations. Users could dial often-called telephone numbers using abbreviated speed dialing codes. These PBXs began to introduce system administration functionality that permitted users to better manage their communication networks.

This new competition in the PBX marketplace simulated customer requests for new and enhanced features. Customers also wanted better system management and maintenance capabilities. The competition between vendors began to focus on feature lists, time to order and install, response to maintenance needs and price. Very quickly this new generation of PBXs provided over 100 features and the vendors found that product differentiation on the basis of features alone was difficult. They started to concentrate on delivery and service. Finally, price wars broke out in an attempt to capture market share. The vendors hoped that aftermarket sales would bring profits. PBX systems became a commodity, and price was a major factor in most purchasing decisions. Most vendors had difficulty in making a profit.

Because the PBX marketplace looked so attractive in the late 1970s, there were almost 40 vendors selling PBX products in the United States. Very few actually made a profit. The number of PBX manufacturers has dropped to slightly over 20 today. In addition, only three have double-digit market shares. Today, most of the analog-installed base, which helped fuel digital sales in the 1970s and early 1980s, has been replaced. And, so far, there has not been a new generation of PBX system to stimulate the market. As a result, many potential purchasers today assume that almost any PBX will meet their needs, and price is the prime consideration. While this is partially true, this book hopes to illustrate significant differences in architecture, feature operation, ease of administration, and maintenance features in current PBX products. Users should carefully weigh these differences when making a purchasing decision. In addition, you must understand whether a particular system will be able to expand and grow to meet unexpected future needs.

It is very likely that in the next few years we will see a new generation of PBX system based on new technologies such as Integrated Services Digital Network (ISDN) protocols. Such systems will be much easier to install and administer. In addition, they will provide more flexibility in the architecture so units may be mixed and matched as needed. They will also be much less expensive. Even before such systems become available it is important to realize that the average price of a PBX has been dropping at a compound rate of about 7 percent per year. All this means that the planning horizon for a PBX should be kept to a minimum. If a PBX has been in place for three years it may be possible to buy a new one for about the same price as upgrading the old one.

The Current PBX Market

At present, when measured by the number of lines shipped, the new PBX market is growing at a rate of about 2 percent per year. The North American Telecommunications Association (NATA) reported in its "1991 Market Review and Forecast" that total annual sales of PBX lines have grown from 3.8 million lines in 1984 to 4.2 million in 1989. That represents 25,000 PBX systems valued at $2.8 billion shipped in 1984 and 29,000 systems valued at $2.9 billion shipped in 1989. By 1995, sales of PBX lines should hit 5.4 million with system sales of 32,000 valued at $3.4 billion. It is the author's opinion that the number of systems projected is probably about right but that the estimate of sales volume is probably optimistic. It is very likely that the sales volume will have difficulty breaking the $3 billion mark.

One additional reason for this flat market is the comparatively rapid growth in the Centrex market. Centrex services are reasonably comparable to PBX services, and the equipment is owned by the local telephone company and is housed on their premises. Therefore, the customer is much less involved in the details of service provision than they would be if they purchased a PBX system. This appeals to many. Others feel that they need more control over their premises communications services. In any case, NATA claims that customer costs for Centrex lines are roughly equivalent to those of PBX lines. While the cost and services are similar, there are several significant differences. Most new PBX systems are digital and support digital telephones. Most Centrex telephones and the lines connecting them to the central office are analog. This will not change significantly until ISDN service is available from the telephone company. PBX systems usually support many more features than Centrex. A PBX user has more direct control over the system. On the other hand, a PBX owner must provide floor space, electricity, and air conditioning for the switch. It must be insured and it is a large capital investment. In spite of these differences customers have been buying Centrex service in such quantities that the Centrex growth rate is much higher than that of PBX systems. The installed base of central office Centrex lines increased from 22 million lines in 1984 to 25 million in 1989. By 1995 the number of Centrex lines should grow to 29 million.

Since there is little growth in the new PBX system market, many vendors have concentrated on aftermarket sales of upgrades, maintenance stocks, service agreements, and peripheral devices and equipment. Most PBX systems are proprietary. That is, only parts from the original vendor may be used for upgrades and additions. Add-on and upgrade equipment is usually much more expensive than the original purchase. Since this is the case, it may pay to buy all of the equipment needed for the next two or three years at the time of initial purchase.

The following table shows how the PBX market distribution has changed in recent years due to regulatory changes and divestiture.

	Market share (%)			
Sales channel	1976	1980	1984	1988
AT&T	74	45	24	26
Interconnect companies	14	37	45	26
Independent telephone companies	10	8	7	5
Manufacturers	2	10	19	31
RBOC	—	—	5	10

The introduction of competition and the deregulation of the PBX market had an extremely negative impact on AT&T's market share. In the 1960s, AT&T's share was in the 90 percent range. After the Hush-a-Phone and Carterfone decisions, interconnect companies, which were usually small sales and service companies for a single vendor's product, proliferated and captured significant market share. However, they suffered from a lack of sufficient resources to withstand the continuing price wars and the movement by manufacturers to establish direct sales channels.

The Regional Bell Operating Companies (RBOCs), who were the distribution arm of AT&T until the end of 1983, are currently prohibited from developing and manufacturing customer premises equipment by the Bell System divestiture agreement. They do, however, function as a distribution channel for PBX products from several manufacturers through separate subsidiaries. They have petitioned the federal court with jurisdiction over the divestiture agreement to overturn the prohibition. Thus far, the court has declined to grant these petitions. Some industry observers believe that Congress will soon transfer divestiture oversight to the FCC. The FCC is inclined to allow the RBOCs entry into the manufacturing arena. If this happens, the market share may dramatically change again since the RBOCs are ideally positioned with end users to be the PBX vendor of choice.

Of the many manufacturers who initially entered the PBX marketplace, only a few have survived. The following table presents the approximate North American market share of each of the major vendors based on the number of lines sold in 1992.

PBX Market Share—1992

Supplier	Market share (%)
AT&T	27.0
Northern Telecom	23.0
ROLM	12.0
NEC	7.0
Mitel	6.5
Siemens	5.0
Fujitsu	4.5
InteCom	4.0
Others	11.0

The future of the PBX marketplace is far from clear. Industry observers predict that it will grow at a rate of only 2 to 4 percent per year. However, this assumes little change in technology or architecture in PBX systems. Perhaps a new generation of PBX systems can provide access to voice services and LAN functions over the same building wiring facilities. Or new, more flexible architectures may be based on ISDN. If this happens, a new round of PBX system replacement may occur. On the other hand, several companies are working on local area networks and metropolitan area networks that will handle both voice and data. These systems may lower the demand for the traditional PBX system.

Major Vendor Products

AT&T

The current PBX product line from AT&T consists of the Definity®
Communications Systems Generic 1 and Generic 3. The Definity
Generic 3 was announced in late January 1992 and comes in two versions. The G3i is ISDN-compatible and supports up to 1600 lines. The
processor will support up to 10,000 busy hour call completions (BHCC).
AT&T still insists on rating processors in busy hour call completions
rather than busy hour call attempts (BHCA) used by almost everyone
else in the industry. Typically, busy hour call attempts include a percentage of calls not completed for one reason or another, such as mis-
dials, busy, and so on. Calls not completed require less processing time
than completed calls. Therefore, a BHCA rating may be typically as
much as 20 to 25 percent higher than the equivalent BHCC rating. As
a result, if measured in BHCAs the G3i processor would support 12,000
to 12,500 BHCAs. The other version of the Generic 3 is the G3r. A
reduced instruction set computer (RISC) processor powers the G3r. The
G3r supports up to 10,000 lines and provides up to 100,000 BHCCs.

The basic components used to build the G3 systems consist of the
following:

Processor Port Network (PPN). A processor port network is
required in all systems and contains the switch processor complex,
including the stored program controlled processor that controls the
system. It may also include some port or interface circuits for lines
and trunks.

Expansion Port Network (EPN). An expansion port network contains additional ports or interface circuits and the basic switching
elements to support them.

Center Stage Switch (CSS). A center stage switch provides connections between the other switching elements contained in the port
networks to form larger systems in version G3r.

The smallest basic system consists of a single PPN. Larger systems
are formed either by directly connecting components or by interconnecting components through a Center Stage Switch. If the system is
composed of one PPN and one EPN, or one PPN and two EPNs, they
may be directly connected together. In larger systems the PPN is connected to the CSS, and the EPNs are also connected to the CSS. Up to
15 EPNs may be connected to the smallest version of the CSS, and up
to 21 EPNs may be connected to the larger version of the CSS.

The processor complex contains a processor, memory, tape drive, an
optional disk drive, and input/output ports to interface with the rest of

the systems and external devices such as a voice mail system or the system management terminals. In the G3i system the processor is an Intel® 80286 processor operating at 8 MHz. The G3r system contains a 32-bit RISC processor operating at 33 MHz. The G3i contains 12 Mbytes of memory while the G3r provides 64 Mbytes.

The switching functions in each port network are provided by a packet bus and a time-division multiplex (TDM) bus. The packet bus provides control interconnectivity. The TDM bus provides 512 time slots. However, 29 of the time slots provide paths for tones, message communications, and features such as music on hold. Therefore, the remaining 483 time slots are available to carry voice and data traffic. For larger systems using the center stage, these busses are connected through the center stage from one port network to another.

There are two cabinet types that house the three basic components: processor port networks, expansion port networks, and center stage switches. A single-carrier cabinet is a self-contained cabinet that houses one carrier. It is 27 inches wide, 22 inches deep, and 20 inches high. These may be stacked up to four high. A carrier, in AT&T terminology, is an enclosed shelf containing vertical slots that hold circuit packs. A multicarrier cabinet is a structure that contains one to five carriers. A multicarrier cabinet is 32 inches wide, 28 inches deep, and 70 inches tall. The cabinets may be provided with either AC or DC power. Either AC or DC power is distributed to each carrier where power converters provide either rectification or DC-to-DC conversion to the proper voltages needed for the carrier. Power backup that provides holdover for short outages from 10 seconds to 10 minutes is provided in the form of batteries contained in the power distribution unit. Uninterruptable power systems are also available as an option.

For increased reliability these systems may be purchased with critical components duplicated. In this case, the processor complex and the center stage switch, if used, are fully duplicated. In addition, the control and information paths from the processor complex and the center stage switch to the expansion port networks are also duplicated.

The software architecture of the G3 systems is divided into an operating system layer and an applications layer. The operating system is Oryx/Pecos, a proprietary real-time system which supports multiprocessing applications with message passing between processes. The system is based on the UNIX® operating system with real-time extensions. Drivers are provided for interfacing to the switching network, mass storage, and other peripheral devices.

The applications layer is composed of three major subsystems: call processing, maintenance, and system management (administration). The call processing program specifies the sequence of actions needed to connect, disconnect, and manage voice and data calls. The management function includes terminal handling, resource management, call sequencing control, and routing. Terminal handling manages the telephone sets along with their feature buttons, displays, and data modules. It also manages the various trunks connected to the system. Resource management involves managing dual tone receivers, time slots, tone generators, and internal software records. Call sequencing control provides the sequencing logic that takes a call in progress from one state to another. For example, setting up a conference call is controlled by the call sequencing control function. Routing controls the selection of the endpoint or set of endpoints of a call. This includes hunting, bridging, call coverage, and least-cost routing.

The system maintenance software performs the functions necessary to provide a high level of service availability. It provides fault detection and system recovery actions. Alarm and error logs are created by the maintenance software for later use by diagnostic technicians. System management software controls the internal processes to install, administer, and maintain the system. A customer or technician may install, test, rearrange, and change equipment and services, and select user and system options. The four main functions provided by the system management software are: measurement collection and reporting, maintenance testing and reporting, translation data backup, and translation database management.

The existing Definity G1 and G2 systems may be upgraded to a G3 system. In some cases this is reasonably inexpensive, consisting of a software upgrade and the addition of circuit packs. In other cases the upgrade may be relatively expensive, requiring the complete change-out of the processor complex in addition to software upgrades and hardware additions.

Northern Telecom

The Meridian® 1 Communication Systems from Northern Telecom are a family of digital voice- and data-switching systems that span a line range of a few lines up to 60,000 lines. They represent the merger of the functionality of the earlier Meridian SL-1, Meridian SL-100, and Meridian SuperNode systems into a single, modular product line.

The Meridian 1 system options 21, 51, 61, and 71 are based on the Meridian SL-1 architecture and provide service in the 30- to 10,000-line sizes. The hardware is based on a modular system called a Universal Equipment Module (UEM). Each of these modules contain all the hardware required to support a specific system function such as a Central Processing Unit (CPU), Network components, or Peripheral Equipment (PE). These modules are essentially self-contained card cages that provide a backplane, power supply, and cabling for the appropriate circuit packs. The UEMs are then stacked on top of one another to form a column. Each column may contain up to four UEMs. At the base of each column is a pedestal that houses cooling fans, air filters, a power distribution assembly, and a System Monitor circuit which provides general system-monitoring capabilities. The top of each column is provided with a top cap assembly which consists of two air exhaust grills and a thermal sensor assembly.

The system is designed so that there are no restrictions as a result of power or thermal constraints. Any circuit card may be placed in any slot, and all modules can be filled to capacity with any logically valid combination of cards. This makes engineering or rearranging a system much easier than might otherwise be the case. The chapter on architecture will make it clear that this universal capability is not always provided. Some universal equipment systems actually have several complex restrictions on the circuit packs that may be used. Some of these restrictions are due to power limitations and others are due to wiring constraints. The Meridian 1 systems provide AC-powered and traditional DC-powered system options.

Call processing, maintenance, and administration of Meridian 1 systems are controlled by software programs stored either as firmware programs, as software programs resident in system memory, or as nonresident programs on disk. Basic functions to manipulate data and control input/output operations, error diagnostics, and recovery routines are "hard-wired" instructions stored in Programmable Read Only Memory (PROM) or firmware. Instruction sequences that control call processing, peripheral equipment, and many administration and maintenance functions are software programs stored in system memory. These programs are interpreted by the firmware programs into machine instructions. In addition to the programs that provide the operational functionality of the system, each system needs a set of data describing the characteristics of the system in terms of configuration and call-dependent information. Generically, in the past, this information has been called translation information and call store information, respectively. In Meridian 1 systems both types of data are referred to as office data.

In addition to the programs and data resident in system memory mentioned above, there are a number of nonresident programs which normally reside on disk. These are sometimes referred to as overlay programs and, when loaded, are stored in the "overlay area" of system memory when required to perform specific tasks. Only one overlay program may be loaded at a time and is removed from the overlay area when no longer required. Overlay programs can be loaded automatically, under program control, or manually, via an administrative terminal. Only one administrative terminal can input into the overlay area at a time. However, several terminals may receive output simultaneously. Overlay programs provide the system interface for maintenance, service change (administration), and traffic measurements.

As outlined in the architecture chapter, every PBX including the Meridian 1 line consists of five types of equipment: the processor system, the switching network, interface equipment, terminal equipment, and power equipment. In the Meridian 1 line these are called Common Equipment (CE), Network Equipment (NE), Peripheral Equipment (PE), Terminal Equipment, and Power Equipment, respectively.

The Common Equipment consists of one or more Central Processing Units (CPUs), memory circuits, and mass-storage devices which control the operation of the system. These units communicate with each other over a common control bus which carries the flow of program instructions and data. The CPU provides the computing power essential for system operation. The system memory stores all operating software programs and translation data including feature information, class of service information, and the quantity and types of terminals. The mass storage unit provides high-speed loading of the operating programs and data into memory. It is used as a backup device for the memory system and to store the nonresident programs.

While the Meridian 1 system option 21 uses a 16-bit processor, all other Meridian 1 system options use a 24-bit processor with a maximum of 16 million words of memory. The mass storage consists of two 3.5-inch, 1.2-Mbyte floppy disk drives and a 20-Mbyte hard disk drive.

The Network Equipment consists of network circuit cards which perform the digital switching of voice and data signals, and peripheral signaling cards which perform scanning and signal distribution. The network equipment interfaces with the peripheral equipment via digital multiplexed loops. A loop is a bidirectional path between network equipment and peripheral equipment for voice, data, and signalling information. Meridian 1 systems provide two types of loops, the standard network loop and the Superloop. Loops are provided in groups. A standard group provides 30 loops while a Superloop provides 120 loops. Each loop is capable of carrying one voice or data conversation but in common Northern Telecom usage a loop usually refers to the entire group of either 30 or 120 loops.

Loops may be assigned to peripheral equipment in a flexible way to accommodate various traffic conditions. In single-loop mode, one peripheral equipment module is connected to one network loop, yielding a maximum concentration of 160 terminations (terminals or trunks) to 30 loops or time slots. In dual-loop mode, half of the PE cards in a module use one loop, and the other half use another loop, yielding two maximum concentrations of 80 terminations to 30 time slots. The Superloop provides 120 time slots to the peripheral equipment. This larger pool of time slots increases the network traffic-handling capacity by 25 percent for each set of 120 time slots. That is, a Superloop is 25 percent more efficient than four regular loops of 30 time slots each. When Superloops are used, the associated Intelligent Peripheral Equipment Modules are divided into segments of four card slots each. There are then four segments on each card shelf. A Superloop may be assigned from one to eight Intelligent Peripheral Equipment segments.

When a Superloop is assigned to a single segment of peripheral equipment, there are a maximum of 128 voice or data ports or terminations in the segment, implying that the configuration is virtually nonblocking with 120 time slots provided for 128 terminations. If two segments are assigned to a Superloop it would be possible to have 120 time slots serving 256 terminations. However, this configuration is usually used when the peripheral equipment is serving digital telephones that do not require data connections or when serving analog telephones. In this case, there are 120 time slots for a maximum, once again, of 128 terminations. Even if half of the telephones require data connections, there are 120 time slots for 196 terminations. If four segments are assigned to a Superloop and no data connections are required, there are 120 time slots for up to 256 terminations. Once again, if half of the telephones require data connections, then the maximum number of terminations is 384 on the 120 time slots provided. Finally, if eight segments are assigned to a Superloop and half of the telephones require data connections, there are 120 time slots for 768 terminations. It is clear that there is a great deal of flexibility in the assignment of loops or time slots to support peripheral equipment. However, this also implies that either a standard, virtually nonblocking configuration is always used, or some amount of engineering effort will have to be provided to configure the system. Also, in the cases with higher blocking probabilities, some amount of thought and planning will have to be provided for all rearrangement and change activities.

The Peripheral Equipment consists of peripheral controller cards which provide the timing and control sequences for peripheral circuits, analog and digital line and trunk cards and circuit card which provide multiplexed digital trunk interfaces and the ISDN Primary Rate Access interface. Digital line cards provide for 16 terminals which may be equipped for voice only or for voice and data. Analog line cards provide for 16 analog telephones each. Each peripheral equipment module or shelf provides 16 slots to contain circuit packs. Peripheral equipment may be remotely located at a distance of up to 70 miles from the main equipment by converting the multiplexed loop signals to a form compatible with the commonly used T1 digital transmission system. In this case, any medium conforming to the standard T1 format may be used, including digital microwave and fiber optic transmission systems.

The terminal equipment supported by the Meridian 1 systems includes standard analog telephones and a portfolio of digital telephone sets. The digital sets operate on a single pair of wires using a technique known as Time Compression Multiplexing to allow digital transmission in both directions. This technique is fully described in the architecture chapter. There are single-line digital telephones and telephones with 8 to 16 programmable line/feature keys. Expansion modules are available to expand the number of line/feature keys up to a maximum of 60. In addition, some models provide space for an optional data module to support data connections to personal computers or terminals. Also available is a telephone with a built-in liquid crystal display. There is even one model, the M3000, that provides a touch-sensitive liquid crystal display that provides access to many features such as a customized directory of more than 250 dial-by-name entries.

The power equipment provides for either AC-powered or DC-powered system options. In either case, –48 volts DC is distributed to each universal equipment module. Each module then contains the appropriate DC-to-DC converters to power the equipment contained in the module. Power backup provision is made with a battery bank in the case of the DC-powered systems, or an Uninterruptible Power Supply (UPS), consisting of a rectifier and an inverter, along with the battery bank in the case of the AC-powered systems.

There are several system options in the Meridian 1 family. System option 21 consists of a single universal equipment module containing a single CPU, slots for network loops, and 10 slots for peripheral equipment circuit packs. The memory size is 768 kbytes. The system can support up to 6 Superloops. Growth to 800 lines is possible by adding up to three more equipment modules to form a column which is four equipment modules high. It is also available is an AC-powered version, but this version is limited to a single equipment module and consequently to 320 ports.

System option 51 consists of a module containing a CPU and half network group functions plus up to three more equipment modules for peripheral equipment. The memory size is 768 kbytes. The maximum configuration is a column of four equipment modules that support up to 1000 ports. The system supports up to 16 network loops.

System option 61 provides a duplicated processor system housed in two equipment modules with each one also containing a half network group. The memory is also fully redundant and consists of 768 kbytes. Equipment modules may be added to support peripheral equipment up to a maximum of 2000 ports. In this configuration the system consists of one column of four equipment modules and one column of two equipment modules. The system will support up to 32 network loops.

System option 71 provides a duplicated CPU and memory system each in its own equipment module. The memory size is 1.5 Mbytes. The system will support up to 10,000 ports in five columns, each of which is four equipment modules high. Up to 160 network loops can be equipped. In this case there are really five sets of 32 network loops each. Each set is interconnected to each of the other sets with up to eight one-way junctors. Each junctor provides thirty channels so that there are up to 240 channels from each set of network loops to each other set.

ROLM

The 9750 Business Communications System (BCS) from ROLM is a family of digital voice- and data-switching systems that span a line range of 50 up to 20,000 lines. Individual systems share a similar architecture and range, from the 9751 Computerized Branch Exchange (CBX) Model 10 to the 9751 CBX Model 70. Each model can be field-upgraded by adding hardware and software. Existing 9751 CBX systems upgrade easily to the latest enhanced models by substituting a limited number of hardware modules and system software.

The Model 10, the smallest of the 9751 CBX family, supports 50 to 600 lines. The hardware is housed in modular, stackable cabinets which allow easy growth. The single cabinet (called modules or shelves in other systems) version is 34" wide, 19" deep, and about 20" high. Two additional cabinets can be stacked on top of the first one, creating a system about 56" high. The first cabinet contains the Computer Common Control (CCC) or processor complex and, in addition, provides 13 slots into which may be plugged line or trunk interface circuit packs. Each of the other two cabinets provides 24 slots for interface circuits. In the Model 10 any interface card can be placed in any slot.

The Model 10 has two power supply options. Both are AC-powered, with a 120-VAC option for single cabinet systems, and 208/240 VAC option for a single cabinet system and required for a 2- or 3-cabinet system. There are rectifiers in each cabinet so that the failure of one affects only a single cabinet.

The Computer Common Control in the Model 10 resides in the first seven slots of the first cabinet. This processor complex includes the processor itself, memory, a floppy drive, and a hard disk. The processor is a 32-bit Motorola 68030 microprocessor with 12 Mbytes of memory. The associated disk systems include one 3.5-inch, 1.44-Mbyte floppy drive and a 3.5-inch, 30-Mbyte hard disk. Also associated with the processor is the System Monitor Input/Output Card (SMIOC). This card provides input/output ports, alarm detection, power and fan failure detection, and internal DC voltage monitoring. The system administration terminals are connected to the SMIOC.

The ROLMbus 295E provides the switching network function in the Model 10. It is a 16-bit parallel bus with a bandwidth of 295 megabits per second. It provides for up to 1728 time slots in the system. Of these, some are used for control purposes, resulting in a maximum of 405 simultaneous two-way connections for voice and data traffic. Each shelf, or cabinet in the case of the Model 10, has an intrashelf bus that operates at 74 megabits per second and feeds voice or data information from the shelf to the 295E bus and back from the 295E bus to the cards on the shelf.

There are several station and trunk interface circuit packs provided for the Model 10. The ROLMlink interface circuit pack provide 16 ports for voice, data, or combined voice and data. They are used to interface the ROLMphone family of digital telephones, other ROLM desktop products, data communications modules, and the PhoneMail voice mail system. These terminals are connected using a single pair of wires using a balance hybrid technique which is described fully in the chapter on architecture. Analog telephones are also supported with 8 ports per circuit pack. Central office, direct inward dial, and 4-wire tie trunks are supported on cards that house either 4 or 8 trunks each. The T1 multiplexed digital trunk interface is contained on a single circuit pack.

Models 40 and 50 are single-node systems consisting of one to five equipment cabinets which measure 32 inches deep, 29 inches wide, and 75 inches high. The Model 70 is a multinode system. Each node consists of one to five connected cabinets of the same size as those for the Model 40 and 50 systems. There can be up to 15 nodes in a Model 70 system.

Model 40 supports all ROLM applications up to 3500 lines. Model 50 also supports all ROLM applications up to 3500 lines, but in addition provides redundancy of all major components for maximum reliability. The redundant components include: the common control hardware, the ROLMbus 295E, shared electronics, and the auxiliary processors. Redundant power can be provided optionally on each shelf as needed. Model 70 supports all ROLM applications up to 20,000 lines and provides the same redundancy as Model 50. Because these models are very similar, only Model 70 will be discussed in detail.

The power supplies for these Models 40, 50, and 70 are distributed and located behind the shelves. In models with redundant power, no interruption of service occurs if one of the power supplies fails. If the system is nonredundant, a power supply failure affects only the shelves served by that power source. Each of these models has two system power options: either –48 volts direct current (DC) or 208/240 volts alternating current (AC). The input power is either AC or DC within a node. However, in multinode systems some nodes may have DC power while others have AC power. In AC-powered nodes, a line conditioning module (LCM) in the bottom of each cabinet converts the AC to DC and distributes –48 volts DC to the power converters located behind the shelves. In DC-powered nodes, the –48 volts DC is distributed directly to the shelves through a battery interface module (BIM) located in the bottom of each cabinet.

The Computer Common Control (CCC) for Models 40, 50, and 70 is located in shelf 2 of the first cabinet in each node. In the fully redundant models (50 and 70), the fully redundant processor system is housed on a single shelf. Each CCC is made up of the processor, memory, Time Division Multiplex (TDM) control, a floppy disk drive, a hard disk drive, a peripheral device controller, the SMIOC serial port interface, a system monitor, and in Model 70 a control packet network interface (CPNI). The control packet network is used to interconnect the processor complex from each node of a multinode system to the processor complex in each other node.

The processors are 32-bit, high-speed, virtual storage 68030 microprocessors which allow each node to support from 18,000 to 30,000 Busy Hour Call Attempts (BHCA). In a redundant system, the primary processor controlling the system is the active processor; the other processor is in a standby mode. The active processor continually transfers new information, as well as calls in progress information to the standby processor. The standby processor always contains current information regarding the state of the system in the event of a switchover from the active processor.

The random access memory (RAM) stores system operating software (the program), system configuration parameters, and operating data (calls-in-progress information). Each model has 12 megabytes of memory standard; another 12 megabytes may be optionally added as required. Battery backup is provided to maintain the RAM memory contents for 12 minutes during primary power failures. If the power is restored within 12 minutes, the system resumes operation without the need to reload the memory. If power is not restored within 12 minutes, the program and system configuration parameters are automatically reloaded into memory from the hard disk upon return of power. Each processor complex contains a 3.5-inch, 1.44-megabyte floppy disk drive and one 5.25-inch, 70-megabyte hard disk.

As in the Model 10, the ROLMbus 295E provides the switching network function in Models 40, 50, and 70. It is a 16-bit parallel bus with a bandwidth of 295 megabits per second. It provides for up to 2304 time slots in each node. These time slots provide a maximum of 1045 simultaneous two-way connections for voice and data traffic in each node. Each shelf has an intrashelf bus that operates at 74 megabits per second and feeds voice or data information from the shelf to the 295E bus and back from the 295E bus to the cards on the shelf.

The station and trunk interface circuit packs provided for Models 40, 50, and 70 are the same as those for Model 10. The ROLMlink interface circuit pack provides 16 ports for voice, data, or combined voice and data. They are used to interface the ROLMphone family of digital telephones, other ROLM desktop products, data communications modules, and the PhoneMail voice mail system. These terminals are connected using a single pair of wires using a balance hybrid technique which is described fully in the chapter on architecture. Analog telephones are also supported with 8 ports per circuit pack. Central office, direct inward dial, and 4-wire tie trunks are supported on cards that house either 4 or 8 trunks each. The T1 multiplexed digital trunk interface is contained on a single circuit pack.

In addition to the switching network functions mentioned above for each of the Models 40, 50, and 70, the Model 70 is provided with internode links (INLs) to carry the voice and data traffic from the 295E bus in one node to each of the other nodes. There can be up to 15 nodes in a Model 70 system. The INLs provide a bandwidth of up to 295 megabits per second in multiples of 74 megabits per second. They are fully redundant and are implemented with twin axial cable for distances of up to 200 feet. Fiber optic cables are used for longer distances of up to 20,000 feet.

Also in Model 70 there is a Control Packet Network II (CPN II) which provides direct processor-to-processor communication between nodes. Multinode operation is based on a true peer-to-peer multiprocessor architecture. The intelligence of the system is distributed throughout the independent processing nodes. The CPN II architecture is a token-based local area network which operates at a speed of 2.5 megabits per second. It is provided on a twin axial cable for distances of up to 1000 feet. At longer distances the CPN II is multiplexed onto the fiber optic cables with the INL information. The CPN II is fully redundant.

If it is necessary to locate one of the nodes of a Model 70 more than 20,000 feet from the others, a remote node connection may be provided. This node can reside at a distance of up to 50 miles from the others. In this case, the transmission facilities between the remote node and each of the others are provided by Extended Digital Interties (XDI), each of which provide 24 two-way communication channels at the T1 facility rate of 1.544 Mbps. As a result they will operate on any transmission line that supports T1, such as microwave spans, infrared links, fiber optic cable, and copper cable. To extend the CPN II to the remote node,

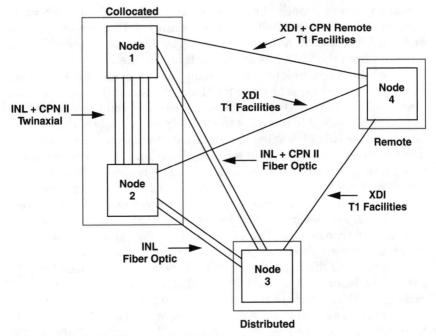

Figure 1.7   ROLM Model 70 node-to-node connections.

a 64-kilobit-per-second channel is used on one of the XDIs. Typical Model 70 node-to-node connections are shown in Fig. 1.7.

The system software for the ROLM product line is written in the C programming language which has all of the advantages of a high-level language, but also allows the programmer easy access to the underlying 9030A processor architecture. It permits enhancement and extension of existing applications, without discarding existing software. It is well suited to the real-time call processing required of modern telecommunication systems. The software is designed in modules to support the distributed processing architecture of the 9751 family of CBXs. The six major modules or subfunctions are the operating system, system integrity or audit programs, call processing, the front-end processor, the configuration system, and the data-switching software.

In multinode systems the modular software integrates individual nodes into a cooperative operating entity, allowing each node to operate both independently and in conjunction with the other nodes. The decentralized and independent databases provide continuous system operation if a node fails, and continuous node operation if internodal communication is interrupted. Node 1 contains the configuration process software for the entire system. However, a system administrator can log on in any node to change the local database of another node. The control packet network transports all of the configuration information between nodes as required. Diagnostics normally must run in the node where the problem lies. However, a service technician can log on in one node, gain access to the problem node, and run diagnostics in the faulty node from the logon site.

The operating system facilitates task management, queuing of common system resources, and allocation of main system memory for all software. It also provides tools for diagnosing and correcting software errors. It can schedule tasks to run at discrete intervals or allow them to demand immediate service as required. It also supports the input/output devices by providing a file system which makes the physical devices transparent to the other programs. It provides a hierarchical file system with directories and subdirectories and also supports keyed record file access which is necessary for the database management software in the configuration subsystem.

The front-end processor software provides a standard interface to the telephone hardware for such functions as scanning and controlling lines and trunks. It shelters the call-processing programs from the details of the interface hardware. It also controls the interaction of the software with the time division multiplex (TDM) switching network hardware. For complex, low-level functions that must take place to operate a specific piece of hardware, it contains a collection of device drivers that implement the specific steps required to complete a high-level function such as "provide dial tone to this telephone and prepare to collect digits."

The system integrity software is an intelligent diagnostic system that attempts to find system hardware problems prior to their impact on communication service. It detects a failure and, when possible, disables the failed facility and uses another facility until the faulty component can be repaired. It performs self-tests and reports errors to a customer engineer through the remote service processor or the system administration console. The engineer can run diagnostic tests to isolate a suspected problem and, after making repairs, can run tests to verify that the problem is corrected. This eliminates the necessity of waiting until the system integrity software completes its routine tests on the same hardware.

The configuration system software enables the system administrator to tailor such system parameters as extension numbers, class of service codes, types of trunks, and trunk groups to meet the customers' needs. It is a specialized database management system which permits the system administrator to select and change specific software parameters as necessary. These activities can be performed without interruption to service.

The call-processing software implements the voice- and data-switching features and applications. It uses the configuration database to find information necessary to process call requests. The voice call-processing software makes voice connections and provides user features such as call forwarding, speed dialing and conference calling. For each call, a data structure is created to track the progress of the call. The data call-processing software handles data calls in a similar way. However, in addition, the data call-processing software checks the configuration of each data line in order to determine how to handle each call setup request. Typical parameters checked include such things as data rate, parity, and clock configuration for synchronous terminals.

NEC

The flagship PBX product from Nippon Electric Corporation (NEC) is the 2400 PBX system. The system is modular in the sense that cabinets are made up of shelves that stack on top of one another to form a stack about 6 feet high. As a result, the system is economical from 50 lines or so up to the maximum of about 14,000 lines. There are processors at the system level, in each module, and on the interface circuit packs, but these processors are tightly coupled with processor busses so that remoting of a module is not allowed. The system-level processors operate in an N+1 redundancy mode by sharing the processing load among a number of similar processors. The switching is provided by interconnected time slot interchangers and is either nonblocking or essentially nonblocking in all configurations. The interface circuit packs are of about the same circuit density as the average PBX system and plug into universal slots. The system is DC-powered, with the rectifier or batteries being external to the system. There are DC-to-DC converters on each shelf which can be duplicated for reliability purposes.

Mitel

For several years now Mitel has been known for inexpensive, small PBX systems. Most of these have been analog in the under-200-line range. Newer systems from the company are digital and have been extended to larger sizes. Even so, the largest part of Mitel's sales remain in the range of a few hundred lines and under. The low cost has been achieved by excellent hardware design wherein the design is implemented in silicon with few intervening steps. Mitel systems warrant serious consideration by smaller companies.

Others

There are a number of other PBX manufacturers represented in the North American market which have less than a 10 percent market share. Some of these have unique characteristics for one reason or another. For example, InteCom provided a PBX for several years that was economical only in large line sizes because the entire duplicate processor complex and much of the switching network, which was centralized, needed to be purchased for even the smallest number of lines. However, the system provided built-in packet-switching capability and therefore enjoyed popularity among large, sophisticated users. As another example, Ericsson provided a system that not only was modular in groups of 200 lines, but the nodes were interconnected with normal T1 digital facilities allowing the nodes to be remotely located from one another with ease. This required duplicated, centralized databases which could be accessed over signaling channels provided in the T1 transmission facilities. Thus, the system allowed excellent flexibility in configuration and rearrangement.

In addition, there were several startup PBX manufacturers who expected to ride the ring-configured local area network boom. They billed their PBX systems as "fourth generation," which claimed to combine the best of traditional circuit switching with a local area network form of packet switching. Both Ztel and CXC based their PBX architecture on a LAN-like ring structure. Unfortunately, rings are not ideally suited for heavy voice traffic. For one reason or another, both these companies ceased production after only a few years.

Chapter

2

Technology

Introduction

A modern, digital PBX is built from the same electronic components as are computers and other telecommunications devices. These include processors, memory, and custom integrated circuits. The dramatic advances in reducing the size and cost of these components has directly affected the size and cost of the PBX systems constructed from them.

In addition, a PBX is simply another component in a telecommunications network. As such, the capabilities demanded of a PBX system are related to the capabilities of the devices used in the telecommunications process. Personal computers are a good example. Before the advent of personal computers, terminals were sometimes switched through a PBX to a remote host computer. The data rates of these connections were rather low. The average person cannot read text at a rate much faster than 2400 bits per second. A few can read at a rate of about 4800 bits per second. Therefore, transmission at rates of more than 4800 bps are not particularly useful for connecting terminals to host computers. However, the personal computer provides local storage, and the transmissions can be from computer to computer with the intention that the information be read or manipulated at a later time. Since personal computers can handle data from devices such as a hard disk at hundreds of thousands of bits per second (Kbps), the bottleneck for some applications has become the communications media. Therefore, there is a desire to provide communications between personal computers, personal computers to servers, and personal computers to host computers at rate up to several million bits per second (Mbps). Witness the growth of Local Area Networks (LANs) to provide such interconnections.

Digital PBXs, while they currently offer data speeds of up to 64 Kbps, have not kept pace with the desire for higher data speeds in the personal computer era. Consequently, while the number of PBX lines used for data connections has grown, the number of PCs connected to LANs has grown at a much faster rate. With the advent of fiber optic-oriented LANs and Metropolitan Area Networks (MANs), the demand for bandwidth will increase once again. Video and high-resolution, full-motion graphics will be transmitted on these new broadband systems.

Does anyone want to put all their data through a PBX? Would anyone want to put video, CAD/CAM, and high-resolution graphics through a PBX? Not any more than a person buys drill bits because they like to collect drill bits. The person who buys a drill bit wants to make holes of a certain size and they use the most convenient tool to do so. What are some of the desires of a customer that may affect the nature of the PBX? Customers want to:

- reduce cost

- increase performance

- increase convenience of installation and rearrangement

- increase the ease of use

This leads to a desire for a single way to wire each workplace, or perhaps provide the communications needs to the workplace without wires if performance and security issues can be overcome. There is a desire to make moves and changes easy by being able to plug any communications device into a standard outlet. If there is a wide variety of information to be transmitted, the interface must support a wide bandwidth, but a customer should only pay for the bandwidth they use and that bandwidth should be available on demand.

All this implies a standard high-bandwidth capability, with dynamic bandwidth allocation that can accommodate both continuous and bursty transmission. In addition, the customer should pay only for what they use. ISDN is an attempt to provide integrated access for both voice and data services at speeds up to 1.5 to 2 Mbps. This will not currently accommodate many high-speed data applications. Video can only be supported if compressed or if it is of lower quality than standard broadcast transmissions. Broadband ISDN should overcome some of these limitations; however, the standards are not yet in place and currently the fiber optics required to each work location may be too expensive. By the year 2000 and the first decade of the 21st Century this will have changed, of course. In the meantime, one approach is to have separate voice, data, and video networks.

What does this imply for PBX systems? As long as PBXs support only voice or voice and data that fits within voice bandwidths, the PBX will be considered primarily a voice communications tool that will accommodate casual data use. This is how most PBXs are currently viewed by potential users and telecommunications managers. But what if a future PBX were to provide LAN-like speeds for data as well as voice connectivity? Or what if a PBX provided economical fiber optic connections and broadband connectivity? In the first place, such a device would probably be called something other than a PBX. In addition, it would have to be compatible with the management systems being used for data and other broadband transmission.

How will technology affect future PBX systems? First we should look at the progress of the applicable technology in general. Then we should look at how the technology affects the hardware and software of the PBX itself. And finally how the progress of other telecommunications systems and devices will affect the demand for enhancements in PBX capability.

The Megatrends of Telecommunications

As mentioned previously, telecommunications is being driven by the same technology trends that are driving the computer industry and other electronic devices. The major driving forces include semiconductor integration, microprocessors, software sophistication, storage capabilities, display techniques, printing techniques, fiber optic transmission techniques, artificial intelligence, and hypermedia.

Semiconductor integration

Since the invention of the transistor in the 1950s, changes in electronic devices have been dramatic. Since telecommunication devices are primarily electronic, they have benefited from this progress. In 1959 the number of transistors that would fit on a chip was one. Now it has surpassed one million. The number of memory bits that can be accommodated on a chip has been growing by a factor of four approximately every four years. As technology limits are reached, the pace is slowing. In spite of this, by the year 2000 there will be integrated circuit chips containing over a billion components.

In addition, there has been significant progress made on new technologies that will extend the current trend for another 10 to 20 years (memory size has increased by a factor of 1000 in the past 20 years). Current semiconductor technology is based on a process called photolithography, a process used to create the components and paths on a semiconductor chip. This process uses visible light in a photographic process to define the components during the manufacturing process. The physical limit to visible light lithography is now on the horizon. The standard etching process will work down to about 0.5 microns or slightly smaller (a micron is one-millionth of a meter) to make computer memory chips having one million transistors. Below that, light waves are too large to render sharp images. X-rays have much shorter wavelengths and can therefore be used to produce much finer detail. The X-ray process can easily etch circuit lines of 0.1 micron (one-thousandth the width of a human hair), which in theory will allow up to 4 billion transistors on a single computer chip.

Medical X-rays are too strong and penetrate the semiconductor material. X-rays produced by particle accelerators known as *synchrotrons* are "soft" or low-energy and work well with the chip-making process. However, until recently synchrotrons have been huge and prohibitively expensive. The key to small synchrotrons is the use of superconducting magnets. The compact synchrotron at Brookhaven National Laboratory on Long Island provides the type of X-rays necessary for the manufacture of ultrafine integrated circuits. The machine began operation in October 1991, costs $32 million, and is 13 feet long by 7 feet wide. It will be used for X-ray lithography in the mid- to late 1990s. IBM has calculated that a single synchrotron would be powerful enough to annually make nearly six times the amount of computer memory used worldwide in 1990.

Microprocessors

Since the early 1970s the performance of microprocessor circuits has improved by a factor of more than 10,000. This is, of course, in line with the general improvements in the fabrication of integrated circuits mentioned above. The first microprocessor, the Intel 4004, contained about 2100 transistors and ran at a clock speed of 800 KHz. Current products have about one million transistors and run at speeds of about 25 MHz. In just the next few years alone improved devices using conventional photolithography will provide chips with as many as five million transistors running at speeds of up to 50 or 60 MHz.[1] This will represent an improvement of about a factor of 10 in performance. The X-ray lithography techniques described above should provide the potential for an increase of at least another factor of 100 by the end of the century or very early in the first decade of the 21st Century.

Progress over the years has been such that the power of a mainframe has been available in a minicomputer in 10 years and on the desktop in

---

[1] Fred Langa, "The End of Intel's Monopoly?" *BYTE*, January 1991, p. 10.

20 years. The power of current minicomputers should be available on the desktop by the year 2000 and the power of current mainframe computers should be available on the desktop by 2010.

Software sophistication

When stored program PBX systems were first introduced in the 1970s the total memory required to store the program, the translation information, and to provide room to store information about each call was a few thousand words. As recently as the mid-1980s some PBXs provided only about one megaword of memory. However, recently there are several PBXs on the market than have memory space for 16 megawords or more of memory. Not only has there been a significant increase in the number and sophistication of features, but system management functions have become more sophisticated and the software to perform maintenance functions on the systems has grown significantly.

Traditionally, PBX software was written in a low-level language because the processors available at a reasonable price were not very powerful and the system was required to process calls in real time. It is no longer practical to write PBX software in these low-level languages—the software packages are simply too big. In addition, much more powerful processors are available at reasonable prices, making it unnecessary.

Currently, most PBX software is being written in C or a similar high-level language. The trend for the future is to move to an object-oriented language such as C++. This will allow for even more sophisticated software to be written by large teams and still be manageable for maintenance purposes.

Early personal computers had RAM memories of a few kilobytes which was sufficient to run the software of the day. Many of today's personal computer programs require one megabyte of memory or more to run efficiently. Mainframe programs consisting of hundreds of thousands of line of code are not uncommon. This more sophisticated software can manipulate much larger amounts of data, resulting once again in an increased demand for higher bandwidths for transmission and switching.

Whole new classes of software are needed and are being produced. There are a few first attempts at software that help extend the thinking process, for example. These thought processors allow you to "brainstorm, cranking out ideas by the bushel basket."[2] They then provide ways to help you organize your thoughts in various different ways to

---

[2] Don Crabb, "Inspiration at the Year's End," BYTE, December 1990, pp. 105–108.

put it all together. These programs are at the stage where they have not yet been recognized as a new and important category of software. Doesn't most computer software help extend the human thought process? No, most programs today require that the desired end result is pretty well in mind before beginning. To prevent later problems one must plan how to set up a database or a spreadsheet before actually filling in the data. The best of these programs will let you easily make changes if you didn't happen to get the design right in the first place. However, they do very little to help in the thinking process about what the design should be. Ultimately the extension of human thinking may end up being one of the most important applications of computers.

Another area where we humans need help, perhaps even more than coming up with creative ideas, is in retrieving information that we know is buried somewhere in the vast information repositories we possess. Storing an important fact in a very large database of information is a little like putting the Ark into an unmarked box in a vast warehouse as was done at the end of "The Raiders of the Lost Ark." As we become able to access huge amounts of information using optical storage and transmission means as discussed below, the need to be able to efficiently retrieve information will become profoundly important. The need for better methods and not just faster searches will be readily apparent to anyone who has tried to find information on a particular subject in an electronic database of published information. It is all too easy to define search parameters that end up providing thousands of references on completely unrelated subjects or, alternatively, almost nothing. You know it's there. The trick is in asking for it in the right way. Perhaps you need to use a slightly different word or words in combination with other words. The need to search CD-ROM databases is stimulating work on retrieval software that can be easily used by the casual user who knows approximately what he or she is looking for.

Storage capabilities

Laser storage techniques are increasing storage densities by enormous factors and reducing the cost per megabyte proportionately. There are basically three types of optical storage: read-only, write-once read-many, and the fully flexible read-write systems. The most popular optical read-only technology is the CD-ROM.

The CD-ROM disk is a laminated structure made mostly of plastic. It is 4.72 inches (120 mm) in diameter and weighs about 0.7 ounces. Data is stored in a three-mile-long track of silvered pits. The silvered surface is supported by a protective undercoating and is sealed from the environment on the top by a transparent upper surface. The resulting disk in unaffected by magnetic fields and is rather rugged although it can be physically damaged. The most serious physical problems result from scratches or dirt on the transparent surface protecting the silvered pits. In addition, damage that causes the disk to become unbalanced, or causes a deformity of the center drive hole, can render the disk unusable.

The data on a CD-ROM is read by a laser beam directed at the silvered pits through the transparent protective surface. The difference in reflection from the pits and the flat surface of the disk where there are no pits provides information to a light sensor which is part of the laser readout mechanism. There is no physical contact with the disk and therefore no wear or degradation except for the physical contact that the drive makes with the center hole to turn the disk. The recovery system works very well; the expected error rate in recovering data is the lowest of any storage medium. To accomplish this, great care has gone into coding the data redundantly so that errors that might occur in reading individual bits may be corrected. One may get an appreciation for the problem of reading the data on the disk by imagining that the disk is enlarged to the size of a baseball field (about 1000 times enlargement). The pits in the three-mile-long track (which is now enlarged to 3000 miles) would be about the size of grains of rice.

The two billion pits on a CD-ROM can carry at least 550 megabytes of digital data. This is enough storage capacity for 250 large books and is the equivalent of about 1200 standard 5.25-inch floppy disks. In addition, personal computer peripheral manufacturers have just introduced CD-ROM "jukeboxes." Typically these hold six CD-ROM and change them as required to obtain the information requested. These units are about the size of a "somewhat long shoebox."[3] Thus this small device intended to be used with personal computers can store about 3 gigabytes or the equivalent of over 7000 5.25-inch floppies.

The write-once read-many or WORM optical technology provides the capability for the user to write on the disk and to read the results as many times as desired. At one time this type of technology was introduced because it was easier to develop than the fully erasable optical disk. However, now that erasable technology is readily available the WORM approach will be used primarily for archiving and for those situations where a permanent record of all data is important.

Erasable optical disks are now readily available at a reasonable price and can be used in much the same manner as magnetic drives. However, the optical disks do not require the extremely clean environment that most magnetic hard-disk drives do. Therefore many optical

---

[3] Jerry Pournelle, "Jukebox Computing," *BYTE*, January 1991, pp. 73–88.

drives will use removable media. They are currently available in 12-inch, 5.25-inch, and 3.5-inch formats. The 12-inch disks hold about one gigabyte of information per disk side or about two gigabytes of information if both sides of the disk are used. The 5.25-inch disks when formatted usually hold about 600 megabytes of information, and the 3.5-inch versions hold about 128 megabytes of information. The 3.5-inch cartridge will easily fit in a shirt pocket.

The jukebox concept is also being used to make storage devices for mainframe computers. In this case, 12-inch optical technology is used; the resulting storage devices have capacities in the *tera*bit range (a million megabits). These tremendous storage capacities cause a demand for higher transmission rates. Simply transferring the contents of a single CD-ROM at modem rates of 2400 bps would take almost three days. To make the transfer of large databases possible, much higher transmission and switching rates are necessary. For example, if it is desired to transmit the contents of a single CD-ROM in two minutes or less, a data rate of at least 5 Mbps is needed.

"Researchers at Bellcore (Livingston, NJ) have developed a new laser-based system that represents a breakthrough in using holograms as computer memory and holds promise for dramatically faster information access. The researchers have built a laser semiconductor array for retrieving holographic images, stored on a glass crystal, at speeds up to one gigahertz."[4] A laser array is used to retrieve holographic images from a photorefractive crystal made from lithium niobate and gallium arsenide. "A single crystal, measuring one centimeter on a side, can [theoretically] store 10 million 'pages of information,' each page containing 100,000 bits."[5] This represents a total capacity of one trillion bits. Each page can be retrieved in less than a nanosecond.

Display techniques

Only a few years ago the standard screen for a terminal or personal computer displayed only 25 lines of 80-column text. Only a single set of characters could be generated, sometimes only in uppercase. Those displays that were more sophisticated could provide lowercase, underlining, and reverse video. Today many personal displays are bitmapped. That is, each dot position is separately addressable. In addition, each dot can take on 256 or more shades of gray or colors in many cases. Newer displays have as many as 1024 by 768 dot positions and can support up to 16 million colors for each dot.

---

[4] Nick Baran. "Breakthrough in Holographic Memory Could Transform Data Access," *BYTE*, January 1991, p. 20.

[5] Ibid.

Where it required only about 2000 bytes of memory to store the information that could be displayed on a character-oriented screen, common displays today require about three megabytes of information to characterize the display. Displays with even higher resolutions are available for specialized purposes although they are expensive.

Significant progress is being made in providing higher-resolution displays at a reduced cost. Some of the newer technologies will allow the screens to be flat and require much less power than previous systems. Ultimately, to transmit moving pictures of near photographic slide quality and in full color will require a transmission speed of about 200 Mbps unless the information is compressed. Even compressed images of this quality will require about 100 Mbps.

If we consider a high-resolution CAD/CAM workstation wishing to retrieve a screen of information in one-half second or less, the transmission medium will have to support data transfer rates of about 25 Mbps. These kinds of displays will place increased demand on higher bandwidth transmission and switching.

Printing techniques

As with display techniques, only a few years ago most printers associated with computers printed characters. Some printed characters faster than others but most printed only characters. Recently the trend has been toward printing everything as graphics, including characters as a set of dots on the page. This allows for anything to be printed which is within the resolution capability of the printer. Many printers print at 100 dots per inch or less. These are currently considered to be of only acceptable quality and it is evident to the reader that they are a computer printout. Printers capable of about 200 dots per inch have been defined as near-letter quality. Most personal laser printers currently print at 300 dots per inch. This is considered to be letter quality and is respectable for graphics. However, commercial quality printing is done at either 1250 dots per inch or 2500 dots per inch. The difference is quite noticeable for reproduction of photographs or computer graphics where tones are to gradually blend into one another.

Just as there is little need to include frequencies above 20 KHz in a high-fidelity system, there is little need to go beyond 2500 dots per inch in printing, because the human eye cannot distinguish any increase in quality. While the current generation of personal laser printers are usually 300-dots-per-inch devices, the next generation has already begun to appear in the marketplace. These printers are reasonably priced and provide output at 600 dots per inch making the output hard to distinguish from commercial quality printing. Since initially higher resolution devices tend to be expensive, they tend also to be shared. In the near future we may see printing devices with resolutions of 1000 dots per inch or higher being shared by groups or workers. At 1000 dots per inch and 24 bits of color or grey-scale information per bit, a standard 8½ by 11 inch page could require as much as 250 Mbytes of information to describe it. Of course most pages will not have nearly this much information on them. However, if we assume a page contains 10 percent of this amount of information and if it is desired to send this information to a shared printer in, say 10 seconds, a transmission speed of 20 Mbps or more is required.

The enormous amount of data that can be displayed, printed, and stored by desktop computers in the near future will substantially increase the need for higher bandwidth connections.

Fiber optic transmission techniques

Just as optical technology is revolutionizing storage and printing technology, it is also revolutionizing transmission technology. The current standard transmission rate through PBXs and the rate specified in ISDN for the B-channels is 64 Kbps. This rate is derived from the standard transmission rate of digitally coded voice signals. Even this is a big improvement over modem data rates of 1200 to 9600 bps. However, given the very large amounts of data that it may be desirable to transmit in the future when very large storage and processing capabilities are available, even 64 Kbps may seem slow. The table below gives transmission times for various amounts of information at 64 Kbps and at 150 Mbps, the lowest rate recognized as standard in the current standard optical digital hierarchies, and 2.4 Gbps, the highest rate specified in the current standards until recently. The latter two speeds bracket the current nominal range of optical transmission speeds. As can be seen, a page at 64 Kbps can be transmitted rather quickly but even a report of 25 pages or so takes almost 10 seconds. At 150 Mbps an entire encyclopedia can be transmitted in less than 10 seconds. The contents of a full CD-ROM could be transmitted in about 30 seconds. The same CD-ROM transmission would require only 2 seconds at 2.4 Gbps.

Transmission Time

Trans. speed (bps)	64K	150M	2.4G
Page	.3 sec	.13 msec	8 μsec
Report	8.8 sec	4 msec	.24 msec
Book	1.5 min	40 msec	2.4 msec
Dictionary	2.1 h	3.2 sec	0.2 sec
Encyclopedia	4.6 h	6.7 sec	0.5 sec
Local library	102 days	1 h	3.9 min
College library	2.8 yrs	10 h	38 min
Library of Congress	71.3 yrs	1.1 days	16.6 h

Artificial intelligence

The controversy over whether machines can think has been raging for at least 25 years. The current status may be summed up in a statement by Jane Morrill Tazelaar, Senior Editor, State of the Art for BYTE magazine, "Is AI dead? Not yet, but it's either going through the throes of a terminal illness or the agony of childbirth." The object here is not to enflame the protagonists or to answer the question. The object of this section is rather to show that simple extensions of what has already been accomplished in the name of artificial intelligence are likely and will be required to handle the vast amounts of information to which we will all have access in the near future.

In the mid-1960s the proponents of artificial intelligence were able to claim that it could be used to get computers to play a pretty good game of checkers (but a lousy game of chess) and do simple geometry proofs. At the time Marvin Minsky said that the current advances in AI could be compared to a person climbing a tree and claiming that they had just completed the first step in a journey to the moon. Today computers can play a decent game of chess or bridge, can recognize voices if trained or a few key words if untrained, and can make some fairly complicated diagnoses or decisions. The question, of course, is whether any of this constitutes intelligence. Whether or not this is intelligent behavior, these capabilities can be useful to us in managing the large amounts of information to which we will have access due to advances in other areas. Perhaps an expert system could be devised to help us in searching for information. We could describe what it is that we are looking for in rather general terms and the expert system could help direct the search. Or perhaps the pattern recognition and learning capabilities available today could be adapted to find patterns in information that would aid us in our search or in organizing what was found. Information could be classified as to the likelihood of its being pertinent to our needs. Information could be linked together based on key words or phrases it contains (see the next section on hypermedia) or grouped into conceptual clusters. Conceptual clustering involves measuring the amount of similarity between objects and then grouping them into clusters where the members of the cluster have similar characteristics.

Expert Systems may help to "combine the subtlety and flexibility of human expertise with the speed and precision of the computer."[6] This could be useful for anyone who must make decisions based on lots of data inputs, particularly if those decisions must be made in real time.

---

[6] Thomas J. Laffey, "The Real-Time Expert," *BYTE,* January 1991, pp. 259–264.

Some examples of those who might benefit include medical workers, financial traders, air-traffic control, plant operations, and process control and communications network control. "[DEC] currently has more than 50 mission-critical knowledge-processing systems in daily operation, systems that save it about $200 million a year."[7] These systems are used for such things as configuring VAX computer systems, routing trucks, and planning the manufacturing process for assembling printed circuit boards.

Hypermedia

Using hypertext methods, it is possible to create "documents" that cannot be printed, at least not in a really useful form. These documents can exist usefully only in a computer. For example, suppose someone created a hypertext document describing an Integrated Services Digital Network (ISDN). The first thing to appear might be a single paragraph with a high-level description of ISDN. In that description it might mention that the user interfaces to ISDN are limited to a small number and that one of them is called the Basic Rate Interface (BRI). By selecting the term BRI the reader is presented with a longer description of the BRI interface in which a function called *passive bus* is mentioned. By selecting passive bus a more detailed description of its operation is presented and so on. Thus a person can read about ISDN at whatever level of detail is desired. If a person is interested they may request additional information on any subject traversing down several levels of detail. Today the subjects on which there is more detail, and the depth of the information, are limited by what the "writer" chooses to include. However, with access to large databases in the future it might be possible to branch into a standard hypertext database so that a person could continue their search. For example, when the investigator above who is looking into ISDN gets to the shape of the electrical pulses defined for the BRI interface, he or she may want to know why it is better to use alternating pulses for the zeros and represent the ones as no voltage. In this case, it might be possible to lead the investigator to a standard electrical engineering hypertext module that explains the tradeoffs between bandwidth, noise, and pulse shape. Ultimately, the investigator might end up at a set of basic definitions provided by a standard set of "dictionaries." One may as well branch out into entirely different subjects. It would be useful to provide a "map" of the way back to the original line of investigation and a set of shortcuts to get back to significant places on the path of investigation.

---

[7] Martin Heller, "AI in Practice," *BYTE,* January 1991, pp. 267–278.

The large databases necessary to provide the backup detail for this type of investigation might be available on optical storage media or may be accessible from a central storage location via an optical transmission medium. Remember that at optical speeds it is possible to download an entire encyclopedia on a subject in a few seconds. This type of hypertext linking between related subjects would make research into new areas much more efficient than with the linear research material we use today.

These concepts are not limited to textual material. The material could be graphics, motion pictures, sound, or combinations of media. For example, a person researching birds might be presented with textual material, slow-motion pictures showing that hummingbirds can hover, or the song of a particular species. In the medical field a researcher might be presented with graphic of a human. He or she could select one of the seven major systems of the body for display and could then get more visual detail on parts of that system. For example, if the circulatory system was selected and displayed, the researcher could then select the heart for a more detailed graphic probably accompanied by text or sound if required.

Several years ago the Media Lab at Massachusetts Institute of Technology created a hypermedia document on a laser disk. They videotaped the process of driving down each street in Aspen, Colorado. They included tours through some of the historical buildings. And this was done in the winter, spring, summer, and fall. The segments of moving images were transferred to a video disk. By using a computer to control playback from the disk it is possible to "drive" around Aspen in any season of the year. It is even possible to stop, enter, and tour some of the historic buildings. When a segment is played it shows the process of driving for one block. At the end of the block the user is asked whether they want to go straight ahead, turn right, or turn left. After the selection, the computer directs the video disk player to play the appropriate segment of the video disk. Thus we are able, through hypermedia, to "visit" Aspen without having to actually go there. The disk, if played straight through, will seem to consist of a series of unrelated segments and will not be particularly interesting.

Combined trends

Taken together these trends are putting tremendous pressure on telecommunications networks to provide increased bandwidth and improved flexibility in the use of that bandwidth. In addition, customers wish to pay for only that bandwidth which they actually use at any given time. It has been said that the information to be transmitted will expand to fill any bandwidth provided. The topics discussed briefly above indicate why this is so. High-speed connectivity has traditionally been provided over short distances first. However, when users find it useful to connect host computers together at 50 Mbps and personal computers together at 10 Mbps, they immediately begin to think how useful it would be to connect together these computer clusters or the Local Area Networks which provide connections for their PCs. What can be done in a room or between offices creates a pressure to do it between buildings or across town. What can be done between buildings creates a pressure to do the same thing on a nationwide or worldwide scale.

This trend is being stimulated by the fact that fiber optic transmission is often the least expensive way to provide a transmission path even though the bandwidth of the fiber is not currently needed. The installation of fiber optics in these cases creates a temporary overabundance of bandwidth. Consequently, bandwidth becomes cheaper. This stimulates demand. Applications that were desired but not implemented because of the high cost of bandwidth suddenly become practical. As the cost of bandwidth decreases, the demand increases. Another characteristic of fiber optics is that cost does not increase as fast as the bandwidth provided. Therefore the cost per Mbps goes down significantly as the bandwidth of the transmission facility goes up. There are significant economies of scale. As the demand for bandwith goes up, the cost continues to decline, which in turn further stimulates demand.

Customer premises switching equipment is simply a link in the chain to provide the bandwidth demanded. A switch will be installed in a department, in a building, or on a campus depending on the amount of information to be transmitted and where it is going. For example, a LAN is installed when the information that is to be transmitted is between a group of personal computers and servers that are relatively close together. A LAN would be inappropriate if the majority of the traffic originated from terminal connections to a host on the other side of town. The PBX has, in the past, filled this local switching function for voice and some voice-band data. It is used when there is a significant amount of calling within the customer's premises as well as to the outside. It provides certain features, routing and cost control functions. In the past the PBX has sometimes been used to transmit data because the wiring was already in place, or the terminals or personal computers needed access to many internal computers as well as long-distance facilities. The transmission rates available to the desktop have been limited to those needed for voice or 64 Kbps.

As long as the PBX is considered to be primarily a voice-switching device there is no incentive to provide increased bandwidth to the desktop since increased bandwidth cannot effectively be used for a voice signal. By providing increased bandwidth for voice, we do not talk faster. We might get higher-fidelity voice transmission, but providing increased fidelity is not a major improvement since most people are pretty well satisfied with digital voice transmission at 64 Kbps. Also there is no incentive for a PBX to provide anything other than circuit switching if the target user is a voice user. There is little benefit and a number of difficulties to be gained from packet switching voice. Therefore unless the PBX comes to be perceived as a multiuse vehicle for voice, data, graphics, and video, it will continue to provide only 64-Kbps circuit-switched channels; if this happens the PBX will be essentially useless in providing for the increased demand for bandwidth outlined above.

Therefore, PBXs will either remain in their traditional role of providing circuit switching for voice and diminish in importance over time until the day when another customer premises vehicle will handle voice as well as other information, or the view of what a PBX is and does will have to change. In the latter case, if a customer premises vehicle does appear that evolves from a PBX that provides both circuit and packet switching at much higher bandwidths, it would probably no longer be called a PBX. And it most certainly could not retain the traditional centralized PBX form of architecture.

Consequently, if the PBX is to participate as a customer premises vehicle of the future in other than a voice and incidental data capacity, the architecture must drastically change and it must include functionality that is not primarily to support voice. If this is to be the case, the designers of such systems will have to look at them as providing capability first for nonvoice functions and for voice as a secondary capability since the nonvoice functions require considerably more capability than does voice. It is the author's opinion that the PBX will not make the transition. The designers of PBXs will most likely try to build a better PBX to compete with other PBX makers, but they are not likely to build something radically different. The radically different products will most likely come from someone else. In effect, the PBX makers will continue to improve the "railroad" while others proceed to build "trucks" and "airplanes." The alternative information transport systems, the "trucks" and "airplanes," will be smaller, more flexible, and more personalized than the current PBX systems. Where a PBX is potentially a multimillion dollar investment that is intended to last from 5 to 10 years, its replacements will be bought in smaller segments that represent an investment of a few tens of thousands of dollars or less. These systems will be optimized to serve a department or smaller work group and will then be interconnected to form larger systems and to interface with MANs and WANs.

Effects on PBX architecture

Having predicted that the PBX will eventually be replaced with something else, it should probably be pointed out that this may take 10 years or longer. What will happen to the PBX in the mean time?

First, the price and size of a PBX have been decreasing by one-half every 10 years. This trend will probably accelerate. The underlying technology used in building PBXs certainly supports a faster rate. The only reason that the rate is not currently faster is that customers consider a PBX to be a major and long-term investment. Therefore, they do not want the product they purchased two years ago to be obsolete. The manufacturers are only too willing to cooperate in this. Even so, every few years some manufacturer will build a significantly less expensive PBX because their old line is becoming obsolete and there is no point in building a new system that is not at least close to the state of the art. Therefore the rate of progress in PBXs will tend to accelerate due to technology enhancements and will be moderated by a lack of competition. The less competition, the slower the rate of innovation. The amount of competition in PBXs is decreasing and therefore so is the rate of change. The rate of change is therefore a compromise between the two forces and will probably result in PBXs decreasing in cost and size by about 10 percent per year for the next few years. A corporate planner would do well to assume at least this rate of change in any planning exercise.

Processor changes

Modern PBX processors are built from the same components as general purpose processors. In fact, it is becoming more and more common to use off-the-shelf parts in the design and construction of a PBX processor. It is likely that newer designs will use a standard bus structure, and peripherals and peripheral interfaces from the personal computer world. These would include floppy drives, hard-disk drives, and optical drives for backup and storage of programs. The system management and maintenance interfaces, both remote and local, will connect to standard terminals and personal computers. Storage devices for call detail recording will take advantage of the devices developed to support personal computers and local networks. Thus the PBX processor and the processor complex will follow the trends of the personal computer/workstation industry. Rapid changes in the size, cost, and power of the processors' memory and storage devices can be expected.

Switching network changes

Switching networks are built from memory chips and others very similar to those used in building computers. Not only would we expect the cost and size of the switching network to follow these trends, it should also be anticipated that it will be more and more likely, as cost decreases, that the switching network will be built to a standard non-blocking configuration. As pointed out above, the switching network is probably not the ultimate source of blocking. However, this move to nonblocking will eliminate one source that must be taken into account. If PBXs are to keep up with the demand to handle information other than voice, the switching network will also have to provide for wider circuit switched bandwidths and packet switching capabilities. It will be necessary to support frame relay in the very near future and cell relay in the next 10 years. Both of these switching functions will be new to most PBX architectures.

Interface circuit changes

Interface circuits in PBX systems consist primarily of the line circuits and the trunk circuits. As such, they are more specialized in their function than most of the other parts of a PBX. Nevertheless, since most of these circuits are currently digital in nature they will follow somewhat the same price and size trends as the technology from which they are built. Manufacturers build custom large-scale integrated circuits to support their proprietary digital telephone sets. With the advent of ISDN many manufacturers will build integrated circuits to support ISDN digital telephones and trunk interfaces. This will drive the cost down at a more rapid rate. In addition, it will allow the customer to eventually buy ISDN digital telephones from any number of manufacturers and not just the one that built the PBX. This will drive the cost of the telephones down as well.

Thus the initial cost of the line circuits will drop and the cost of telephones will drop. However, to help to retain their margins we can expect to see the cost of additional and replacement circuit packs to remain high, since once the PBX has been purchased only the line circuits made by the manufacturer of the PBX will work in that PBX. These trends must be taken into account by any telecommunications manager planning on purchasing a new PBX.

The platform PBX

Recently PBX vendors have borrowed a term from the world of computing and have proclaimed their product to be a *platform* PBX. This is meant to imply that the PBX can be enhanced and upgraded in the future, a form of investment protection. The term may also be used to mean that they are providing a hardware/firmware platform that, while performing no function by itself, can be used as a foundation for a wide variety of applications. Furthermore, it is implied not only that the software may be changed and upgraded without affecting the hardware but that the opposite is also true. That is, the hardware may be upgraded without affecting the software. Until recently this latter feat has been a rare occurrence. In most PBX systems the software has been so tightly coupled to the hardware that almost any hardware change required corresponding changes in software. This has changed recently with the adoption of true operating systems running on the PBX processors and the use of commercial upgradable processor families as the processors themselves.

It is now possible, in some PBX lines, to upgrade the processor without changing either the software or the rest of the hardware. For example, the Fujitsu 9600 originally offered an 80286 processor rated at 55,000 Busy Hour Call Attempts (BHCA). It is now possible to upgrade the 80286 processor to an 80386 processor rated at 110,000 BHCA. The trend with platform PBXs should be to allow the upgrade of any major component of the system without having to change the other components or the software. This will require that all of the major interfaces between these components be standardized, at least within the design organizations of the vendor company. This raises an interesting question. If these interfaces are standardized anyway, why not publish them and allow other vendors to provide specialized components for parts of the system?

The ISDN BRI interface will have the effect of standardizing the digital telephone to PBX interface. Many manufacturers are standardizing, or at least proposing their version as a standard and publishing the specifications, the PBX processor to host computer interface. These latter interfaces have been dubbed "Request and Status Links" and have been the subject of hot debate recently. What remains to be standardized? First, the command and control connections between the line and trunk circuit packs and the processor. Second, the status and control links between the processor and the switching network. And finally the digital data paths between the line and trunk circuits and the switching network. By standardizing the two paths from the line and trunk circuits into the rest of the system along with the circuit pack size, the powering, the connector, and the pin-outs, multiple vendors could produce plug-in circuit packs for the PBX system. If this seems to be too much to believe, it may simply be because of the traditional proprietary nature of PBX systems. Compare the last suggestion above to the bus in an open personal computer. The specifications on card sizes, the connector, and the pin-outs are known along with the powering arrangements. This allows multiple vendors to build a variety of cards that will plug into the PC. Many of these cards are used to interface with peripherals which are not from the same vendor as the maker of the PC.

In any case, it has been suggested that a platform PBX must include the following attributes:

- A switching network that uses Time Division Multiplexing (TDM) and PCM
- An upgradable central processor
- A decentralized processor architecture
- A high-level programming language
- Uniform software across all models
- Open interfaces

The question is, of course, if this is the definition of a platform PBX, is it sufficient to ensure the goals sought in creating the platform in the first place? Will these attributes provide for a long life of the system? Will it be able to be upgraded and changed as needs change? Will it protect your investment for a reasonable period of time?

If the switching network is put together using standard TDM/PCM techniques it means that the switch is a circuit switch based on the capability to switch 64-Kbps chunks at a time or perhaps multiples of 64 Kbps. This in turn implies that it will switch voice quite well except that it would be nice now that 32-Kbps Adaptive Differential PCM has been standardized if it would switch in 32-Kbps chunks or, even better, in 8-Kbps chunks. However, this is still circuit switching and is inefficient for many kinds of data. Packet switching should be incorporated if the switch is expected to handle bursty data. In fact, if the switch is to be fully ISDN compatible it should provide packet handlers for not only standard X.25 packet switching but also a frame handler to support frame relay. Of course it is possible and allowed by the standards to circuit switch a data terminal to a packet network. But this means that the transmission facilities from the PBX to the packet node will be inefficiently used when compared to the case where the switch provides packet- and/or frame-handling functions.

In addition, the standard TDM/PCM switching fabric is not prepared to handle broadband rates. It will be a rare PBX that will switch T1 rates of 1.544 Mbps in the next few years. Broadband rates are usually defined as beginning at 50+ Mbps. Thus if the platform PBX is based on standard TDM/PCM it will remain basically a voice switch that also handles low-speed occasional data needs. What could be done to prevent being boxed in by the switching network? The best approach would be to define a standard high-level interface between the processor complex and the switching network which has a message set which is rich enough and/or expandable so that new fabrics may be added and controlled by the existing processor complex.

A decentralized processor architecture and an upgradable central processor is certainly desirable and would be a real asset in the battle against obsolescence. However, the amount of processor power required may be easily underestimated. The rich set of features made available by ISDN may require a processor that is considerably more powerful than current models just to handle the same number of BHCA. If broadband capabilities are added the possible combinations of features, options, and bandwidth expand dramatically. In addition, communicating with and allowing partial control to a host attached by an RSL link may require additional processing capacity simply because many new feature combinations are possible. A higher percentage of digital telephones will also demand more processor horsepower. Taken together, these advances may end up requiring a processor several times more powerful than the current model. This can be accomplished only by using a processor family that has several upgrades but also one where continued development is likely to lead to additional family members that will support existing software while significantly increasing the capacity of the system.

Higher-level programming languages are also desirable. However, they usually are somewhat less efficient than lower-level code and therefore require more powerful processors. The same is true of a standard operating system. A standard operating system is a good idea but once again one must be prepared to pay for it in processor horsepower. What is really required is the use of an Object Oriented Programming language. This too will require more processor horsepower; however, it is needed to simplify the software design so that it is practical to design and maintain the very large software systems required by modern switching systems. This is particularly true when the design is to provide for the external control of the system by host computers over RSL connections. To make this dual control system work where the PBX processor handles the low-level processing to control the details of the system and the host handles the high-level call handling decisions, the software modules (objects) must be very well defined and well behaved. Each side must understand exactly what will happen if a certain command is issued. If this is not the case, unexpected things will happen frequently and troubleshooting will be extremely difficult.

What is key to a long-lived, extensible PBX system is not only open interfaces, but interfaces that are standards. This will allow the individual parts to be upgraded not only by the original manufacturer, but if they don't provide needed capabilities in a timely fashion, others will. There is, however, a long way to go before PBX manufacturers will be convinced that anything but the most obvious interfaces should become standardized. The suggestion that a PBX manufacturer should build a PBX with a standard bus into which may be plugged circuit cards from multiple vendors would be met with dismay. This is not something that anyone even thinks of doing. Yet the open architecture of the original IBM PC had a profound effect on the industry which was not all bad for IBM. In any case, a telecommunications manager wishing to ensure a long life for the next PBX which they purchase should insist that as many of the interfaces as possible in the PBX follow open standards.

Capabilities for the next decade

Figure 2.1, adapted from a diagram created by Bell Communications Research (Bellcore), indicates the broad range of potential traffic that future PBX systems may be called upon to carry if they are to serve as more than just voice switches. Bandwidths required vary from a few bits per second to nearly a gigabit per second and session time ranges from parts of a second to several hours. What is not shown in Fig. 2.1 is the wide disparity in the nature of the transmissions. Some of these transmissions are nearly continuous with very little traffic in the reverse direction, such as high-speed computer file transfer. Some transmissions are almost continuous and have nearly equal forward and backward transmissions such as a voice conversation. Some of the transmissions are very bursty where the actual transmission takes place for 10 percent or less of the connect time, as is the case when an interactive terminal is connected to a host computer. Some transmissions may periodically require huge bandwidths for only short periods, such as a CAD/CAM workstation loading a drawing from a server. Other transmissions simply require huge bandwidths such as high-definition TV. These are all potential candidates for transmission on a customer's premises and for interconnection to a wide-area network. Today's PBX will not be adequate for many of these applications. A system available tomorrow that would handle all or a majority of these applications would probably not be called a PBX. And yet there is a need for systems to handle these requirements. Is it necessary or even desirable that a single system provide for all of these applications? Not necessarily.

Switching capability has always lagged behind transmission capability. However, we are on the brink of the age of fiber optic transmission at speeds of hundreds of megabits per second to multigigabits per second. When the most economical way to serve a workstation is to provide it with a pair of fiber optics, is there any longer any need to provide a pair of wires for voice or a coaxial cable for data? Given the right standards and interface protocols the fiber has sufficient capacity to handle almost any combination of information needed. Switches will have to follow. It may be that different switching fabrics are needed to

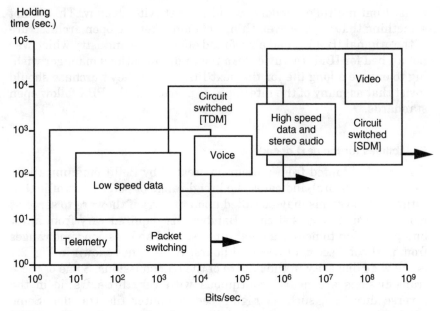

Figure 2.1 Dynamic range of services.

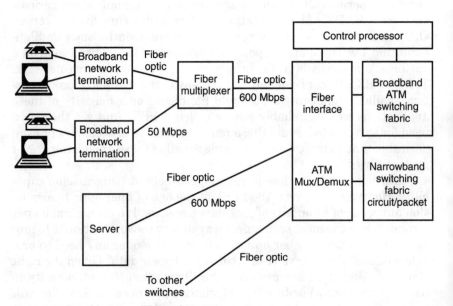

Figure 2.2 Possible future PBX architecture.

provide for voice, video, and data. However, all of these should be multiplexed over the same fiber, and the control of the switching should be done from one processor complex or a group of processors interconnected to provide seamless applications to the user. Such a system would have the architectural configuration shown in Fig. 2.2. It can be seen that this system has many of the characteristics of today's PBXs and also resembles a LAN with servers.

Architecture/Design

Introduction

Ultimately, your objective in selecting and installing a PBX system is that it perform the intended communication functions at a reasonable cost and that it does so reliably. The range of objectives can be extensive. Some want a PBX to perform only basic telephone communications functions at the lowest possible cost. The PBX purchase and operation is viewed as a cost of doing business. Others expect the PBX to perform communications functions that will give the enterprise a competitive advantage. In this case the PBX is viewed as a strategic resource which should pay for itself through increased revenues. In either case, very few people feel the need to have a detailed knowledge of the inner workings, design, and architecture of the PBX system or network which they install.

However, whether your needs fall at one of these extremes or somewhere in the middle you will find it difficult to predict the ability of the system to handle increased load or to adapt to new functions without some knowledge of system design and architecture. One approach is to rely on the manufacturer for this information. Another is to hire a consultant to analyze whether the systems will fit your future needs.

A typical approach is to issue a statement of need, either formally as a Request for Proposal (RFP) or informally, perhaps only a verbal description to a sales representative. The PBX supplier then responds that they can either meet your requirements or they cannot, but they have some other alternative which would meet your ultimate need. If your request simply states what you need now, you will have little or no idea of how the system might be expanded or changed to meet your future needs. The usual remedy for this is to try to predict future needs in the original request. However, it is surprisingly difficult to precisely define a present need let alone try to describe several potential future scenarios.

At this point many future PBX owners will hire a consultant to listen to their needs and write the RFP. This is a reasonable thing to do since the consultant is probably familiar with many similar situations, and knows what questions to ask and how to describe the requirements in terms that the vendors will find familiar. However, if the consultant writes the RFP, including your best guess at future needs, and then evaluates the responses, making a suggestion as to the proper course of action, you may have no greater understanding of the trade-offs and alternatives than before. Once the system is installed and the consultant has gone and you need to determine the feasibility of growth or changes, you must once again rely on the manufacturer or recall the consultant.

A common solution to the problem of future growth and changes is for the customer to request a system that will handle all possible future situations. In many cases a consultant will recommend a system with extreme flexibility so that there will never be a situation where the system does not meet the customer's future needs. In most cases much of the potential capability of the system is never used. One example of possible overspecification is in the area of network blocking. The safest thing to do is to specify nonblocking. If the network never blocks, you will not have to worry about getting into a future situation where the PBX is causing a problem because it is the communications bottleneck. There may be situations where this is appropriate. However, if the RFP requests nonblocking, then all of the responses will be configured and priced that way. You will never know if one of the vendors could have provided a very good blocking characteristic (say, one connection in a million blocked) for significantly less money. Perhaps all of the vendors could have reduced their cost significantly for a one-in-a-million blocking level, in which case you may end up paying a substantial premium for a capability that may not really be needed. Another possibility is that the price from one of the vendors may be essentially the same whether the system is truly nonblocking or whether it is only close. In this case, if that vendor also meets all of your other needs, the nonblocking capability may be almost free. In this case, you may feel comfortable in paying for it even if it may not be needed in the future. It is cheap insurance.

One approach to this dilemma is to ask for many different alternatives in your original request. In this case, ask for a system configured for three or more levels of blocking, say one in a few hundred, one in a few hundred thousand, and strictly nonblocking. The typical result might be: one vendor is a little less expensive for nonblocking but significantly higher than the other two for the other two blocking conditions. One vendor gives a price for all three cases but complains that you just missed one of their standard design configurations and that if you had specified one in one hundred thousand rather than one in five hundred thousand their price for that configuration would have been significantly lower. The third vendor's price is very attractive for the one-in-a-few-hundred case but higher for the other two cases. This analysis is now rather difficult. If you stick with the nonblocking specification, vendor A is the least expensive. However, you could save a considerable amount of money by going with vendor B if you accepted one in a thousand rather than one in five hundred thousand as the appropriate middle level of blocking. Finally, you could save even more money by going with vendor C if you are willing to accept blocking of one in a few hundred calls.

Perhaps one way out of the problem is to investigate the potential to save some money now by accepting some level of blocking, but upgrading the system in the future if and when totally nonblocking is needed and justified. So you ask the vendor to give you upgrade costs from the lower levels of blocking to nonblocking. This is relatively easy for vendor A since their systems cost essentially the same in any blocking configuration. However, with both vendors B and C there is a significant cost associated with the upgrade which makes the final cost of nonblocking much more expensive than installing the system that way originally. Finally, since you are specifying a complete system, the costs for the blocking configuration of the network will not be neatly broken out as is implied by this simple example, but they will be buried in the total system cost figures and can only be obtained by looking at the difference in the different models.

Another example may further illustrate the point. Suppose you do not feel that you need ISDN (Integrated Services Digital Network) service now but that you probably will in the future. You could ask the manufacturer if the system can be upgraded to ISDN. The answer will be yes. This answer is not much help since the real answer may be a *qualified* yes. If you don't specifically ask what the qualifiers are they may never get mentioned. You need to ask whether the BRI passive bus is supported, and whether user-to-user information on the D-channel is supported. You should ask if the PBX has a packet handler and provides packet switching, or only provides circuit connections to a packet network. So you ask how much it will cost. You will get the price of circuit packs, telephones, and software. If you stop here you may be surprised when you actually upgrade. You may find that the system will handle significantly fewer ISDN telephones than it will analog or even the manufacturer's proprietary digital telephones. Why is this? Because ISDN messages contain much more information than the signaling from an analog telephone or even a proprietary digital telephone. These may cause the processor to be overloaded or there may be a signaling bottleneck between the line and trunk circuits and the processor due to the architecture of the system.

You should ask the vendors whether the PBX would work if all your lines and trunks were ISDN. You may find that one or more of the systems that were adequate for non-ISDN service will not handle an all-ISDN system. You may find that some are significantly more expensive than others. In short, you have the same problems described above in the nonblocking example. One vendor may be less expensive in a non-ISDN configuration but may not be able to handle the all-ISDN situation. Another may be able to provide all-ISDN but at a significantly increased cost. Once again this may turn out to be a fairly difficult problem even by itself. When combined with several other significant issues, the problem of picking the optimum PBX for your needs becomes quite complex. The only satisfactory solution is to know the basics of the architecture of each system so that you can decide which ones can more easily be expanded to meet potential future needs. You may wish to avoid this by hiring a consultant, however.

Consequently, the only real alternative to ensure that you are able to select PBX equipment that meets your needs now and in the future, and at the same time minimizes the overall cost, is to understand the basics of PBX architecture. This will allow you to make decisions as to what PBX is best, or at least enable you to ask the critical questions the answers to which will reveal what you need to know to make an informed decision. In this way you will not have to rely on vendors' unverified statements and you will be able to get the maximum benefit from a consultant if you feel you need to hire one. For systems larger than a few hundred lines, hiring a consultant is probably a good investment. Even for smaller systems if you have any unusual requirements it may make sense. In either case don't rely on the consultant to make all the decisions. Rather use the consultant as an additional source of expertise and someone who may think to ask some of the questions which you may have forgotten to ask. In any case you should be aware of the reasons why a consultant asks certain questions and understand the implications of the answers. If you don't, you are letting the consultant make the decision for you rather than using their expertise as a resource.

System

A PBX is a made up of several major components, each of which is relatively easy to understand when broken down into the elementary parts. But if a PBX is approached as a single entity, determining what is and is not important about its structure and design is often difficult.

As a result this chapter will provide you with the definitions and an overall model of a modern PBX system that will help you to understand the function and purpose of each of the parts. In addition the appropriate criteria to be used in evaluating each part will be presented.

A system is a collection of diverse physical components organized and interconnected to perform a specific set of functions. For example, the major components of an automobile engine cooling system are its radiator, fan, water pump, thermostat, and hoses. These diverse components are organized and interconnected to perform the functions of circulating water and dissipating excess engine heat.

Architecture

The architecture of a system specifies the characteristics, arrangement, connections, and interactions of its components. For example, the architecture of an engine cooling system specifies:

- the characteristics of its components—the size and shape of the radiator, the diameter of the fan, the number of fan blades, and so on

- the arrangement of its components—the placement of the radiator, the placement of the fan with respect to the radiator, and so on

- how the components are connected—the return hose connects the bottom of the radiator to the water pump, and so on

- how the components interact—the fan shuts off when the thermostat indicates that the water temperature is below 180°F, and so on

Therefore, defining the architecture of a PBX involves identifying its major components, describing the characteristics of these components, and understanding the relationships between these components.

Importance of architecture

The architecture of a PBX is a major factor that influences its ability to perform useful functions in a cost-effective manner. A PBX with a poor architecture may perform well, but may be difficult to modify or may lack flexibility. Alternatively, a PBX with a poor architecture may perform a wide variety of tasks, but none of them very well.

A PBX with superior architecture not only meets its design objectives in a cost-effective manner, but can also be:

- Expanded—to increase in size or capacity. For example, if more call-handling capacity is required, the processor can be replaced with a more powerful version.

- Enhanced—to provide additional functionality related to the original switching features. For example, if a group of agents is required to handle incoming calls to an "800" sales line, automatic call distributor (ACD) functions can be added by installing new software, agent consoles, and a management information system (MIS) computer.

- Updated—to provide newer, more cost-effective technology. For example, if new technology allows twice as many components per circuit pack, these packs can easily be adapted to the system without significant hardware or software changes and with little or no disruption of service.

- Interconnected—with other systems to perform new and enlarged tasks. For example, if a voice mail system is needed, it can be connected so that integrated services are provided. That is, the PBX sends information about the caller, who was called, and the reason for transfer to the voice mail system. The mail system then tells the PBX which message-waiting lamps to light.

The architecture of a system also determines whether it is *open* or *proprietary*. A system with open architecture has interfaces that conform to widely accepted standards that have been created by standards-making bodies, such as the International Telegraph and Telephone Consultative Committee (CCITT) or the American National Standards Institute (ANSI). In theory, such a system can easily be connected to other systems which conform to the same standards, even if those systems are provided by other vendors. A system with proprietary architecture is one which has interfaces that are unique to a single vendor. Typically, such a system can only be connected to other systems from the same vendor.

Software architecture

While it is much less tangible than the hardware, the software used to run a modern PBX system also has an architecture. This architecture is just as important as, and in some ways more important than, the hardware architecture. A superior software architecture allows the software that controls the system to be expanded, enhanced, updated, and interconnected with other software easily, cost-effectively, and in a reasonable period of time in a manner similar to that discussed for the hardware. Vendors seldom discuss the software architecture of a system in any amount of detail. However, you as a customer have a right to know whether the software has a well-thought-out architecture or a poor one which has simply grown beyond its usefulness.

Much of the software which runs a PBX must execute in *real time*. That is, the software must make decisions about the timing of trunk sequences or whether to disconnect ringing during short intervals of time defined by the hardware requirements or by customer expectations. The case of disconnecting ringing is a good example. When a customer answers a ringing telephone they expect the ringing to stop almost immediately. In electromechanical equipment, a relay that sensed the current flow when the telephone was answered was used to immediately disconnect the ringing. However, in modern digital telephones there is no change in current flow to indicate answer. Rather a digital message is sent to the switch indicating that the telephone has gone *off-hook*. This message must be processed and the correct action must be determined by a processor. In this case, the correct action is to cause the alerting function to cease or be turned off. If this process takes too long (say one-half second or more) because the processor is busy doing other things, the customer will be annoyed and probably consider the system to be slow and unresponsive. If the processor that handles the decision to disconnect ringing is the main processor, it must be designed so that the probability is very high that it can get to and complete the task of causing the ringing to be removed in the required time. Of course, this processor is handling many other tasks at the same time.

It should be obvious that the main processor does not have to handle such a request. If, for example, there were a processor associated with each line or small group of lines, it could be programmed to know that when it receives an off-hook message for a ringing line it should remove the ringing and then pass the off-hook message on up to the main processor. This is one form of distributed processing that helps to increase the performance of a PBX processor system. This topic will be discussed in more detail below.

Computer telecommunications architecture

The ability to interconnect software in one processor, say a PBX processor, with that in another processor, say the processor in a voice mail system, requires another form of architecture, a computer telecommunications architecture. Because of the growing need to communicate between computers to form distributed processing networks, these computer communications architectures have been studied in detail and many have been adopted by various standards bodies. The same issues addressed in these standards apply to the communications between processors in a PBX system or between the PBX processor complex and external processors, such as hosts, or adjunct processors such as the voice mail processor described above.

The interprocessor communication within a PBX system need not follow an accepted and published standard because it is all internal to the system. However, if standards are used it is almost certain to provide increased flexibility in terms of being able to remotely locate parts of the system or to be able to substitute a new component that is improved or perhaps even from a different vendor. On the other hand, it is extremely important for any interprocessor communication between the system and any other system to follow a widely accepted standard that is appropriate for the purpose. This will allow you, as a customer, more flexibility in choosing adjunct or host equipment to use in conjunction with your PBX. Until recently most PBX vendors used proprietary communications protocols for these interconnections. Or at best they invented some communications method, published or licensed the protocols, and tried to talk as many computer vendors into supporting it as possible. In many cases an existing computer communication standard or such a standard with some slight extensions could be used to perform the same functions. The latter course is the preferred one since it is likely that many vendors already support the standard or will support the standard, giving you a much wider choice of suppliers. ISDN is one such standard that can be used for many applications. The interface specifications and D-channel signaling protocols are appropriate for many telecommunications-related functions.

Also, over the past several years significant progress has been made on standardizing a model for computer communications by the International Standards Organization (ISO). The model is called the Open Systems Interconnection (OSI) model. When possible, the interprocessor communications associated with PBXs should in the future be based on protocols that conform to the OSI model. The ISDN protocols do so by providing slight extensions to allow one channel to control others in the case of the D-channel and to provide multimedia capability.

An Architectural Framework for the Analysis of Modern PBX Systems

Most often a PBX is described in the manufacturer's literature, a trade press article, or an RFP as one complete entity. There is usually a physical description detailing the number of cabinets and the number of circuit pack carriers within the cabinets. The arrangement in terms of module or node and how many lines each will serve is presented. The interconnection of modules and how many can be connected is mentioned. The processor type and whether it can be duplicated is pointed out. And finally, some call handling and environmental statistics are usually provided. RFPs are much the same, except they specify the number of lines and trunks desired and the statistical measures that the system must meet such as busy hour calling capacity and environmental conditions.

The problem with this approach is that when they are described in this way every PBX is different. It is difficult to compare any particular architectural feature. Each system has good and bad characteristics. Each one had different capacities and capabilities in every area. Worst of all, with this type of description, it is very difficult to perform useful analyses comparing one system to another.

The solution to this problem is to use a structured framework for analyzing the architecture of a PBX. When such a framework is used in conjunction with a consistent analytical methodology, definitive architectural comparisons between modern PBX systems can be made. The next few sections of this chapter provide the architectural model that is needed to develop such an analytical framework. Some important criteria that can be used to evaluate the major components of a PBX system are also presented.

The eight major architectural components of the suggested architectural model include the:

- switching network

- processor system

- adjunct processors

- network interface circuits

- terminals

- power system

- system software

- computer communications interfaces

Figure 3.1 shows the general architectural relationship among the major hardware components. The power system is not shown explicitly, but it provides the necessary power of each of the other hardware components. The system software resides in the processor system and is also not shown explicitly.

The switching network provides the actual circuit switched connections to connect the various ports together. These ports can be interfaced to analog telephones, digital telephones, data terminals, analog or digital trunks, and other equipment through the network interface circuits. The purpose of these circuits is to provide the proper interface needed by each device or facility and to separate the signaling information from the information that is to be switched. The signaling information is directed to the processor system while the voice or data information is directed to a port on the switching network. The processor system consists of one or more processors and peripheral equipment that are used to run the program that controls the system. The

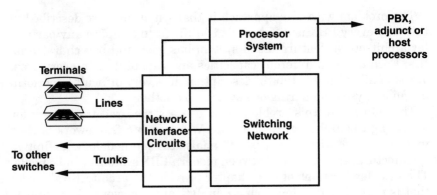

Figure 3.1 The major components of a modern PBX.

system software that resides on this processor complex has a software architecture that is important to the operational characteristics of the system.

As shown in Fig. 3.2, an adjunct processor or host may be connected to the PBX system in two ways. First, there may be switched connections from appropriate interface circuits that deliver information from another port on the system. This allows terminals or network facilities to be connected to the processor. In the case of a voice mail system this would be the connection for the telephones that receive announcements from the system and are able to leave messages. In the case of a host computer it would be the way that the terminals get connected to ports on the host. Second, there may be direct computer communications connections between the PBX processor and the adjunct or host processor. This is a normal computer communications link and should follow a widely used and recognized standard. This link is used to allow the PBX processor and the host processor to exchange information. For example, a voice mail system receives information about a call over this link. It also uses it to indicate to the PBX which message waiting lamps should be lit. In the case of a host processor used as part of an ACD system the host receives the dialed number and the calling number identification over this link. This enables the host to fill the screen of the agent with pertinent information about the caller and the reason for calling as obtained from the dialed number. The computer communications protocol architecture pertains to these links.

Of course, each of these major components are made up of a set of smaller pieces and has an architecture of its own. That is, the switching network can be defined by describing a number of smaller pieces and how they are arranged and connected together to make up the network. The processor complex is similarly made up of several sub-

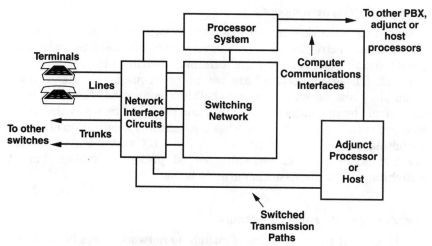

Figure 3.2 The major components of a modern PBX.

components such as processors, memory, storage devices, and communications ports. The functions of each of these major components and the alternative used for the architecture of each will be discussed in more detail below.

Switching Network

Since the primary function of a PBX is to make switched connections between terminals, between terminals and trunks, or to other things connected to the system, the switching network which performs the connection function is arguably the most important component of a PBX. As such it should be the first major component to be examined. Not only is the switching network important to the major function of a PBX, providing switched connections, but it can also positively or negatively affect other parts of the system. For example, a complex network design requires that the processor perform much more work to establish each call, and therefore negatively impacts the ability of the processor to quickly establish calls. In fact, a poor network design can restrict the alternative processor complex designs. For example, certain network architectures may require that the processor be centralized rather than distributed, due to the requirement for the processor to have information from many parts of the network in order to make connection decisions. To allow easy distributed processing, the network must be constructed such that the distributed processor need only know about the state of the network part with which it is associated.

What a switching network does

A switching network is an arrangement of switching elements (metallic contacts, electronics, or time-shared electronics) used to make temporary connections between the terminals or trunks connected to the network. These connections are set up on request and taken down when they are no longer required. While connected, they are used exclusively by the endpoints involved and provide a two-way transmission path of a fixed bandwidth. This is essentially a definition of circuit switching which all PBXs provide. A few PBX systems also provide another major form of switching called packet switching. Packet switching will be discussed separately below.

Overview of a PBX switching network

The functional characteristics of switching networks have been extensively studied. While there are many possible network configurations, the most fundamental switching network design consideration is to provide PBX users with an acceptable grade of service as economically as possible. The cost of the network is evaluated in terms of the number of switching elements required to build the network. The grade of service is measured by the probability of a call attempt being blocked. A call is blocked when a requested connection cannot be made because there are no paths available through the network to an otherwise nonbusy endpoint.

In its most basic form, a switching network can be viewed as a simple array of input/output lines arranged in rows with connecting circuits or links arranged in columns. Figure 3.3 depicts a basic switching network. The X's indicate switching elements which can connect any user to any other user. For example, Fig. 3.4 shows two connections, one between Station A and Station B using the first link on the left, and one between Station C and Station D using the third link from the left. It is easy to see that any input may be connected to any output that is not already in use. This particular switching network is nonblocking because a user can always make a connection to any other nonbusy user regardless of how many connections already exist within the matrix.

In this approach, the number of links required is one-half the number of input/output lines or telephones. The number of switching elements required grows in proportion to the product of the number of lines times the number of links. For example, in Fig. 3.3, there are eight lines and four links. Therefore, the number of switching elements necessary is $N^2/2$, where N is the number of lines. While this is a satisfactory design for small systems (up to a few hundred lines), the ever-increasing number of switching elements makes it prohibitively expensive for large systems.

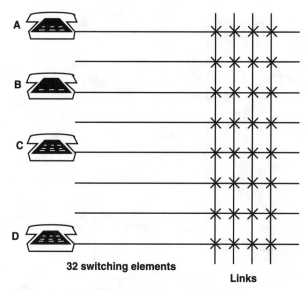

32 switching elements

Links

Figure 3.3   A simple switching matrix—nonblocking.

To avoid this problem, fewer links may be used which in turn requires fewer switching elements, as shown in Fig. 3.5. This matrix requires only 16 switching elements. However, since there are only two links, this switching matrix only handles two calls at a time and is, therefore, no longer nonblocking. If a third user requests service, he or

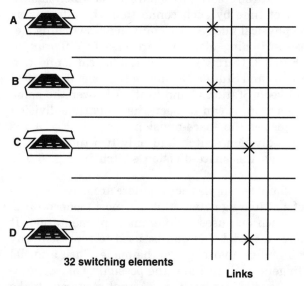

32 switching elements

Links

Figure 3.4   Connections in a simple nonblocking matrix.

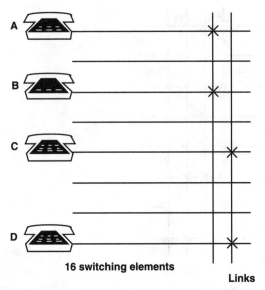

16 switching elements

Links

Figure 3.5   A simple switching matrix—blocking.

she will be blocked. Thus, expensive switching elements have been eliminated at the price of system availability.

The switching networks described above are known as space-division switches. Another way to save valuable switching elements is to "time-share" them, using a technique known as time-division switching. If the links in Fig. 3.3 are replaced by a single bus which is used only for a short period of time by each connection, the effect is as shown in Fig. 3.6. The physical bus requires only eight switching elements, although these switching elements are clearly different in nature than those in Fig. 3.3. During time slot 1, a logical connection is made between Station A and Station B. During time slot 3, a logical connection is made between Station C and Station D over the same physical path. By comparison, it can be seen that this time-division switch is nonblocking, just as the space-division switch in Fig. 3.3 is nonblocking. If the time-division switch had only two time slots, it would be blocking, just as the space-division switch in Fig. 3.5 is blocking.

As illustrated by the simple examples above, there are many ways to approach the design of a switching network. While most modern PBXs have digital networks, each one uses a different implementation. It would be helpful to have a model that can be used to represent any of them so that meaningful comparisons can be made. Such a model is shown in Fig. 3.7. In general, it includes the possibility of a concentrating function and a switching function. Actual systems, to be

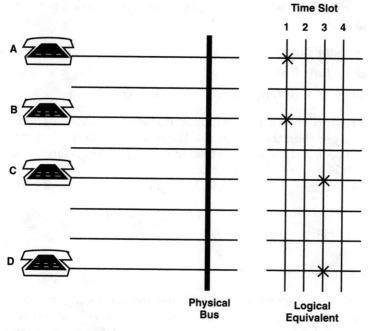

Figure 3.6   A time-division bus.

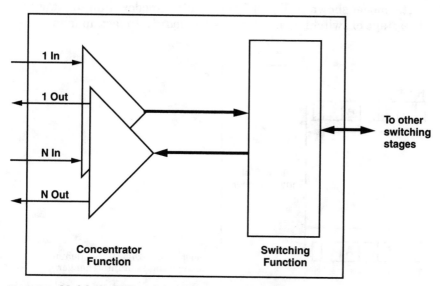

Figure 3.7   Model of a PBX module (sometimes called a node).

described using the model, may not contain both functions. There are three possibilities:

- the concentrator function only
- the switching function only
- both the concentrator function and the switching function

A concentrator connects a set of inputs to a set of outputs where the input set is larger than the output set. It cannot connect one input to another input or one of the outputs to another output. Therefore a concentrator is less general than a switch. Furthermore, in actual PBX systems, the switching function may be implemented in one of three ways:

- an electronic space-division switch
- a time-division bus
- a time slot interchanger (TSI)

A TSI operates on multiplexed digital input streams as shown in Fig. 3.8. An input stream from N sources is shown with two of the sources, A and B, identified. The TSI interchanges the time slot positions or the data units from A and B so that the output stream has the positions of the time slots of the two reversed. The circuitry is quite different, but when the TSI is combined with multiplexing and demultiplexing, the effect is the same as the time-division bus shown in Fig. 3.6.

The model shown in Fig. 3.7 is actually a model of an input/output (I/O) stage of switching usually called a *module*. In fact, in many small

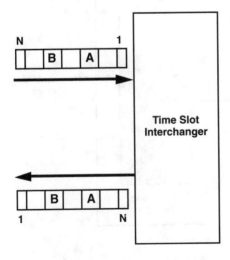

Figure 3.8 Digital switching using a time slot interchanger.

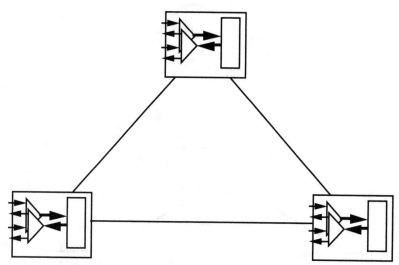

Figure 3.9   Directly connected modules.

PBX systems, the entire switching network consists of a single module. However, for reasons relating either to economics or physics, a single module cannot be expanded to arbitrarily large sizes. The largest useful size is about 1000 lines. Switching networks for larger PBXs are often constructed from a number of modules using one of two methods. In the first method, the modules are connected directly together as shown in Fig. 3.9. This is a reasonable approach for a small number of modules; however, the number of links required grows rapidly as the number of modules increases. This becomes prohibitively expensive. The formula for the number of links is $N(N-1)/2$, where N is the number of modules to be connected. For example, to connect 10 modules, 45 links are required. Therefore, for systems expected to grow larger than about 4 or 5 modules, another method is used. The modules are interconnected with a center stage of switching as shown in Fig. 3.10. The purpose of the center stage is to connect modules together. The center stage can be distinguished from the module because it is not directly connected to inputs and outputs (e.g., telephones) themselves.

In summary, the switching network model used in this book has one or more modules. Each of these modules contains concentrators or a switching function or both. If there is more than one module, they are either directly connected together or are interconnected with a center stage.

The switching networks of all modern PBX systems may be analyzed using this model. Evaluating the architecture of a PBX switching network involves identifying the parts of the actual switching network

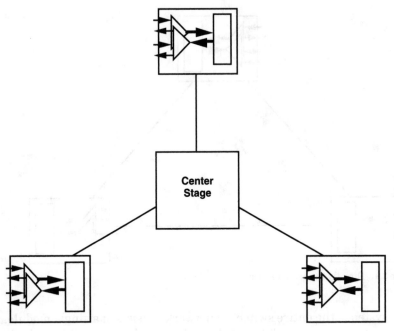

Figure 3.10   Modules connected with a center stage.

according to the model, describing the operation of each part, and describing how the parts are interconnected.

Examples of switching networks

Figure 3.11 shows four examples of switching networks that are implemented in typical PBX systems. For ease of reference, these are labeled Systems A through D. In each case, the modules are shown as rectangles with an "I" in the upper left-hand corner.

In System A, each module contains 24 pairs of concentrators as shown in Fig. 3.12. Each pair of concentrators supports up to 80 inputs and outputs. The concentrators are connected to a network bus through a TSI. The bus consists of a number of multiplexed links, one for each concentrator. Each concentrator transmits on its assigned link. However, each concentrator may receive from any link. Since all of the information on the links is available to any concentrator, connections are made by each concentrator selecting the correct time slot from the appropriate link.

There is no center stage in this system. Each module is connected to every other module with direct links. This scheme is called a *junctored* architecture. Collectively, the direct links are called a junctor group.

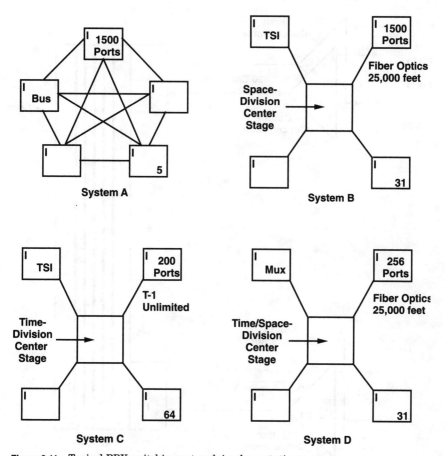

Figure 3.11   Typical PBX switching network implementations.

In this example, the junctor group has up to eight links. Each link has 30 two-way channels.

In the next example, System B has modules that contain concentrators connected to a TSI which handles 512 time slots. Each of the 1500 ports in a module has full access to any of the time slots. The modules are connected to a center stage using fiber optics of up to 25,000 feet in length. The center stage is an electronic space-division switch.

System C has a similar architecture to that of System B. However, it has smaller modules of 200 ports each. The concentrators provide full access to a TSI which has a sufficient number of time slots so that each module is nonblocking. The other major difference is that the limited number of ports allows European-style connections to be made to the center stage using 30-channel 2.048-Mbps digital multiplexed links. In

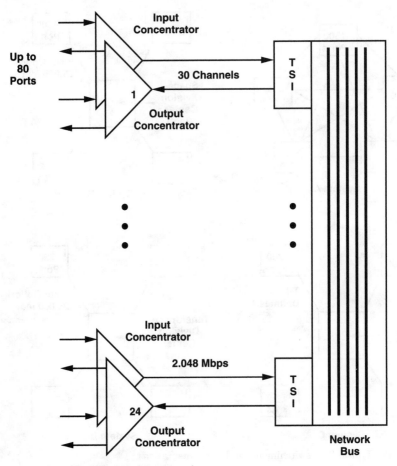

Figure 3.12   The I/O stage for system A.

North America or Japan, each link is limited to 24 channels. In some cases, two 24-channel links must be provided, resulting in some wasted capacity. On the other hand, where the fiber optics in System B are limited to 25,000 feet, these links may be run over arbitrary distances and can make use of leased facilities or private microwave systems. Also, the smaller port size of the System C modules makes them more suitable for small groupings of terminals than the System B modules.

The last example is System D. Here the center stage consists of a number of TSIs connected together with a high-speed bus. Each of the TSIs supports a module over a fiber optic link of up to 25,000 feet in length. The modules are relatively small, with a capacity of 256 lines each. They contain multiplexers only. In this design, there is no switching function in the module. The overall switching network is nonblocking.

Evaluation and comparison criteria

There are several criteria which may be used to evaluate and compare switching network architectures. The most important criteria include:

- blocking characteristics
- maximum port capacity
- ease of growth
- remoting capability
- reliability
- cost

Figure 3.13 is a matrix summarizing the characteristics of the four example systems with respect to each of these criteria.

Blocking characteristics. The blocking characteristics of a switching network can be determined by examining three of its components:

- the module
- the links between modules, or between the modules and the center stage
- the center stage, if any

System	Blocking	Capacity	Growth	Remoting	Reliability	Cost
A	Blocking in mux access. Good between modules.	5000 to 7000 ports	Must add many links in larger sizes	Easy for remote groups, but remote modules are difficult.	Failures confined to small groups	Two or three modules - low cost. Five or more modules - expensive.
B	Full access in modules. Between stages depends on links.	8000 to 30,000 ports	Must add center stage for two modules	Remote groups require special design. Remote modules are easy with fiber.	Center stage and links to modules critical	One module - low cost. Two modules - add center stage. Many modules - low cost.
C	Full access in modules. Between stages depends on links.	12,000 ports	Must add center stage for three modules	Remote groups require special design. Remote modules - unlimited distance	Center stage and links to modules critical	One or two modules - low cost Three modules - add center stage. Many modules - low cost.
D	Full Access in modules. Non-blocking overall.	8000 ports	Must have center stage from the beginning	Remote groups require special design. Remote modules are easy with fiber.	Center stage and links to modules critical	Three or four modules- high cost Many modules - low cost.

Figure 3.13 A comparison of PBX switching network characteristics.

The architecture of System A uses a large number of small concentrators that make the module less efficient than systems which provide full access from each of the ports to all of the time slots. Therefore, given equal numbers of ports and time slots, System A will have a higher probability of blocking than the other systems.

Blocking characteristics are also quite dependent on the capacity of the links between the modules. System A has links of about the same capacity as Systems B and D. However, as System A grows, links are added to every module, which provides added capacity for handling calls between modules. In the other systems—B, C, and D—all of the traffic between modules must travel through the center stage. The capacity of these links does not increase as additional modules are added. Therefore, for systems with a center stage, the capacity of the links from each module to the center stage is critical. Of the center stage examples, only System D provides sufficient link capacity to provide nonblocking capabilities.

Regardless of the blocking characteristics of the links, the center stages are usually nonblocking. However, if they provide space-division switching and no time slot interchanging, then it is possible that matching time slots on the incoming module link and the outgoing link cannot be found. This is called time slot mismatch blocking. It is on the order of 1 in 100,000 and may be negligible unless the system is engineered for nonblocking or close to nonblocking service.

Some telecommunications managers feel that a PBX must be nonblocking if it is used to switch data. However, when a limited number of computer ports are provided for economic reasons, some of the data calls to the computer during the busy periods of the day will be blocked because no computer port is available. The same may be true of trunks to other locations. When this is the case, the PBX switching network only needs to have a probability of blocking that is about ten times less than the blocking generated by these sources to have a negligible effect on the overall blocking calculations.

Maximum port capacity. The capacity of a PBX to support a certain number of ports may be limited by either the system physical size or the architecture of the switching network. This section examines limitations that are caused by the switching network.

In System A the number of direct links to every module grows rapidly with the number of modules. Therefore, System A's total port capacity is limited to about four or five times the module size. The number of links required is $N(N-1)/2$, where N is the number of modules. For any N greater than four or five, this architecture becomes impractical. For example, if the number of modules required was 15, then the number of links necessary to connect them together would be

105. For the other systems, the upper size is limited by the design of the center stage and the size of each module. The size of the modules is limited by its switching function capabilities.

For example, the TSIs used in Systems B and C are limited, due to the speed of the components involved, to providing about 1000 time slots or less. In System B, the TSI has 512 time slots. The maximum number of ports these time slots will support depends on the blocking characteristics desired.

In System C, the 200-port physical capacity of the module has been chosen small enough so that the module is nonblocking. As shown in Fig. 3.11, the maximum number of modules is 31 and 64, respectively. Therefore, the upper size limit of System C is about 12,800 ports, while the upper limit of System B depends on how many ports are used on each module due to blocking considerations. The upper limit could range from about 8000 ports to 30,000 ports.

Ease of growth. This is not a problem when adding the second or third module, but it gets increasingly more cumbersome as the number of modules increases. Systems B, C, and D avoid this problem by employing a center stage. Once the center stage is installed, growth is accomplished by connecting a new module to the center stage. However, installing the center stage itself is a significant effort and may increase the cost of small systems by as much as 50 percent. In System B, when the second module is added, the center stage must be added as well. In System C, two modules are connected together with direct links. However, when the third module is added, the center stage must also be added. In System D, the center stage is required for even the first module, since there is no switching function in a module.

Remoting. There are three common methods for locating and accessing terminals that are remote from the main part of the system. The first method is simply to put some of the terminals on longer wires. These are called extended-range arrangements or off-premises lines. They may require repeaters or amplifiers for each line to extend the signaling range or provide acceptable levels of transmission. In addition, if the wires run outside of a building, protection from foreign voltages, such as those caused by lightning or power crosses, is required for each line. Although this method of remoting is expensive for more than a few stations, it is equally effective for any of the architectures.

The second method is to move some of the line circuits to the remote location. The connection between the line circuits and the terminals remains the same. However, the connection between the line circuits and the rest of the system must be extended. This is usually accomplished by using a digital multiplexed line at 1.544 Mbps (T1) or 2.048

Mbps (European standard). These types of connections operate at distances of at least 100 miles and are suitable for small groups of terminals. There is no local switching in these remote line circuit groups. This method of remoting is easily implemented in System A, since the concentrators are connected to the network bus with a 2.048-Mbps link anyway. The concentrators or multiplexers in the other three systems are too large to be moved away from the switching function. Therefore, if remoting of this type is required, the concentrators or multiplexers must be redesigned so that a part of one may be split off from the remainder of the inputs in the module.

The final method of remoting is to move an entire module to the remote location. For switching networks with junctored architectures, a remote module must still be connected to every other module. Figure 3.14 illustrates this situation for a system with five modules where one is remoted. Thus, module remoting for System A is uneconomical when large numbers of modules are necessary. Module remoting is better suited for Systems B and D. However, in these systems, the modules are connected to the center stage with fiber optics. The distance is limited to 25,000 feet and a right-of-way must be available for the fiber. System C modules may be remoted to indefinite distances because they are connected to the main system on digital multiplexed facilities.

Reliability.   From a reliability standpoint, System A's switching network architecture is very good. Since the concentrators and TSIs come in small segments, a failure on any one of them causes only a limited

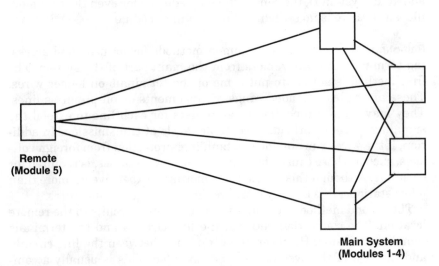

Remote
(Module 5)

Main System
(Modules 1-4)

Figure 3.14   Remoting a module in a junctored switching network.

number of ports to fail. Between modules, the failure of one of the junctors in a junctor group simply reduces the traffic capacity between the two modules involved.

In the other systems, the center stage is a critical element. If it fails completely, no calls can be completed between modules. In fact, in System D, since there is no switching in the modules, a failure of the center stage means that no calls can be completed. For this reason, the center stage in System D is always duplicated and the other center-stage-based systems offer optional redundancy. If a link between the center stage and a module fails, that module cannot communicate with any other module in the system.

Cost.   From a user cost standpoint, System A's network architecture is low cost for small systems. For a single module, System A can be expanded in smaller increments than the other systems. For two or three modules, System A's junctored architecture is less expensive than requiring a center stage. However, as the number of modules grows, the cost rises faster than systems with a center stage.

At the other end of the spectrum, System D has larger multiplexers and requires a center stage that is at least partially equipped even for a single module. Therefore, it is unlikely to be economical at small line sizes. Systems B and C fall in the middle between these two extremes. However, System C has smaller modules and, therefore, may have the edge over System B for small system sizes. System B requires that the center stage be added with the second module. In System C, this does not occur until the third module is added. Adding the center stage causes a significant increase in the cost of Systems B and C.

The Processor System

The second most important component of a PBX is the processor system. In this book, the term processor system refers to all of the processors in a PBX and the interconnections between them. Some of the processors in a PBX are connected to several additional components, such as a backup system or a number of input/output ports. The term *processor complex* is used to describe such a processor along with the additional components that are connected to it.

What a processor does

The processor system provides the intelligence to run the PBX. It controls the switching network; receives information from the interface circuits about requests from, and the status of, the lines and trunks; and provides control and signaling information to the interface circuits. It also collects, stores, and manages information about calls handled by the system as well as potential maintenance problems uncovered during call processing.

Component overview

This section describes two aspects of PBX processor systems. First, the components of a typical processor complex are discussed. Then, a hierarchical model of the multiple levels of processors that are typically used in a modern PBX system is introduced.

The processor complex.   A PBX processor complex may consist of off-the-shelf products designed for general purpose use, or the components may be custom designed expressly for the purpose of call processing. In either case, the architecture is very similar to that of personal computers (PCs) or other small computer systems.

A block diagram of a typical processor complex is shown in Fig. 3.15. This complex is organized around a *processor control bus,* which allows easy communication between components of the complex. Connected to the processor control bus is the system memory. This memory stores information required by the processor in the form of a program, which is a set of instructions used in running the system. System memory also contains data describing the settings for each system parameter. For example, this data might include information about the types, numbers, and locations of trunks connected to the system; the types of routing patterns used; and the features associated with each telephone in the system. This data is called the *translation information,* or just *translations.* Finally, the processor uses part of the system memory to store temporary information about the status of each call. This part of the memory is known as the *call store* or *scratch pad memory.*

Another component connected to the processor control bus is the *backup system.* The backup system reloads the program and translation information into the memory whenever there is a power loss. Some PBXs store backup information on a digital tape cartridge, while others store it on a hard disk. If a hard disk is used, floppy disks are also required to back up the hard disk. The backup systems are also used to load new software into the system. The new software may be provided on tape cartridges or floppy disks.

The remaining connections to the processor control bus are input/output (I/O) functions. The first I/O function is an interface to a terminal or PC used for system management. In some cases, this interface is also used as a maintenance access point.

The next I/O function is the interface to other processors outside of the PBX system. These other processors range from applications pro-

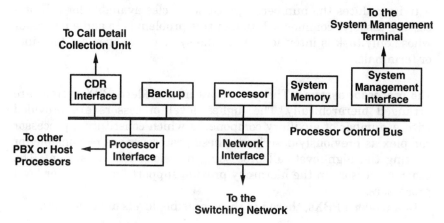

Figure 3.15   A typical PBX processor complex.

cessors that perform auxiliary functions such as directory services or voice mail, or host processors that provide financial applications, to processors on other PBX systems. In the latter case, this interface allows different PBXs to share information so that calls passed between them are handled almost as if they were a single PBX.

Another I/O function is used to store information about the calls completed by the system so that usage charges may be billed back to the appropriate departments. This function is known as *call detail recording* and is usually accomplished by sending information over the call detail recording interface to a temporary storage unit. The storage unit is then periodically polled by a host computer or PC. The information gathered is combined with information from other PBXs in the network and processed to provide the necessary reports.

The final I/O function, usually called the *network interface,* is the information and control channel from the processor complex to the switching network. This channel allows the processor to control the setup or takedown of the connections in the network, receive information about the requests being made by the terminals and trunks, control trunk operation, and operate various displays and lamps on the terminals. Although the switch control function and the information passing function could be separate, they are usually combined into one.

In some cases, the processor complex may also include a special processor associated with an I/O function. For example, the system management interface shown in Fig. 3.15 implies that the processor responsible for call processing must also provide the computing and memory storage functions for the system management terminal. This activity reduces the number of processor cycles available for call processing tasks. A common solution to this problem is to add a processor whose only task is interfacing with the system management terminal or terminals.

The processor system. The processors in a modern PBX system are arranged hierarchically. The highest-level processors are provided with a number of auxiliary components which constitute a processor complex as previously described. These processors are responsible for making the high-level call processing decisions for the system. The other processors in the hierarchy provide support for the higher-level processors.

In a modern PBXs, there are three possible levels of processors:

- System level
- Module level
- Port level

These three levels are shown in Fig. 3.16.

It is quite common, though not necessary, for a PBX to have a system-level processor. If there is a system-level processor, it makes the

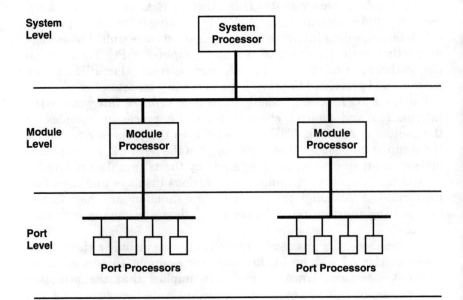

Figure 3.16   A model for analyzing PBX processor systems.

high-level call processing decisions and is part of a processor complex. In some cases, it may be the only processor in the system.

In many PBXs, processors also appear in the individual module, as defined in the previous section on switching networks. If module-level processors are present, each module contains its own processor.

There are two types of module-level processors. The first type supports the system-level processor or processors by handling some of the real-time intensive functions performed in the module. This type of module-level processor scans for lines or trunks requesting service, handles trunk timing sequences, turns lights or ringers in the telephone on and off, and so on. These are low-level functions that require significant processing time. By performing these functions in the module-level processor, the system-level processor is free to handle the higher-level decisions. In this processor configuration, as the system grows, another module-level processor is automatically added to the PBX with each new module.

The second type of module-level processor is used in PBXs where no system-level processor is present. In this case, each module-level processor must not only handle the real-time intensive processing for the module, but must also make the high-level call connection decisions as well. In this configuration, each module-level processor will have many, if not all, of the associated components of a processor complex as shown in Fig. 3.15. To complete calls between modules, a robust communication mechanism, such as a local area network (LAN), is required to interconnect the module-level processors.

Finally, some PBXs include processors on their port boards. Port-level processors are usually much smaller than system-level or module-level processors. They handle the timing functions required by the trunks, turn on and off lights and ringers on digital sets, and report requests for service. Thus, the port-level processors may perform functions that are similar to module-level processors. Except for very small PBXs, port-level processors don't make high-level call completion decisions. Also, it is feasible to have a system with port-level processors and module-level processors, but no system-level processors. The most likely combinations of PBX processors are summarized in Fig. 3.17.

An architectural description of a PBX processor system involves a description of which processors exist in the system, their functions, and how they are connected together. For some purposes, it may also be necessary to further describe the components of the processor complex.

Examples of processor systems

Figure 3.18 shows three examples of processor systems that are implemented in typical PBXs. System A has a system-level processor com-

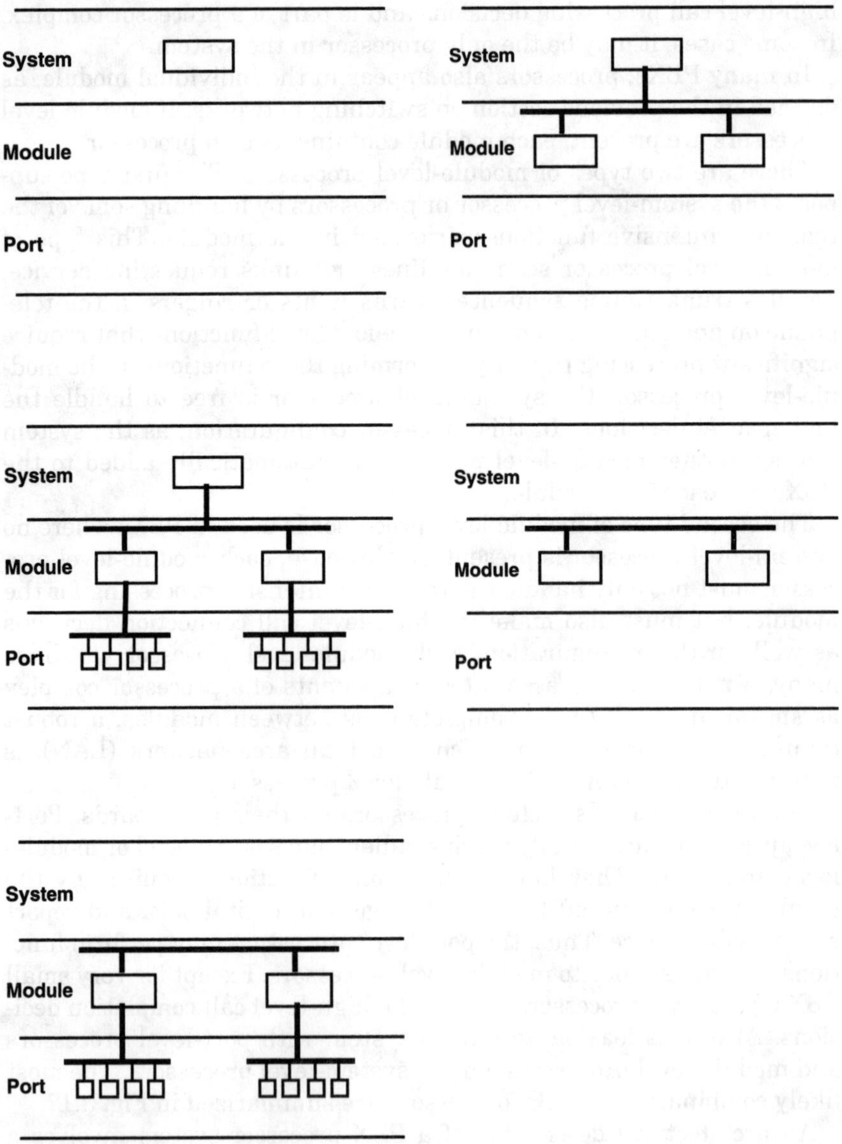

Figure 3.17 PBX processor system alternatives.

plex that is directly connected to a number of module-level processors. These connections carry multiplexed data streams that control the switching connections and communicate with the interface circuits. They use coaxial cable for short distances and fiber optics for longer distances. The system-level processor makes all of the high-level call processing decisions and, therefore, the modules cannot operate independently.

System B has a system-level processor complex as well as module-level and port-level processors that are interconnected through a series of common busses. The system-level processors operate in a load-sharing manner. They make all of the high-level call processing decisions and, therefore, the module-level processors do not operate independently. The port-level processors handle the real-time intensive work such as scanning, digit collection, and timing functions. The module-level processors collect information from the port-level processors and participate in controlling the switching network.

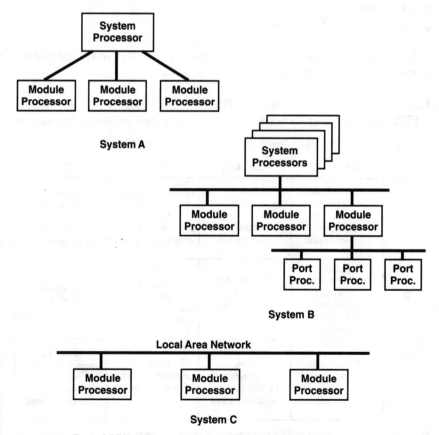

Figure 3.18 Typical PBX processor system implementations.

In System C, there are module-level processors, but no system-level processors. The module-level processors are connected together with a LAN which passes information between them whenever intermodule calls are made. If the LAN fails, the modules will continue to process intramodule calls, but will not process intermodule calls. For longer-distance connections, systems such as System C use a standard packet network instead of a LAN.

Criteria for evaluating and comparing processor systems

There are four major criteria for evaluating PBX processor architectures including:

- busy hour call handling capacity
- remoting capabilities
- reliability
- cost

Figure 3.19 summarizes these criteria with respect to the processor examples described previously.

Busy hour call handling capacity. The most important characteristic of a PBX processor system is its ability to handle call requests during the

System	Capacity	Remoting	Reliability	Cost
A	Large at small line sizes. Grows with modules. Cannot adjust.	Easy, but high data rate requires fiber optics	System Processor critical. Module processor affects only one module.	High for small systems. Low for large systems.
B	Medium at small line sizes. Grows with modules and ports. Adjustable.	No remoting due to bus architecture	System Processor critical. Module processor affects only one module.	Medium for small and large systems.
C	Grows with modules. Cannot adjust.	Depends on LAN. Use of packet switch makes remoting easy.	Module processor affects only one module.	Low for small systems. High for large systems.

Figure 3.19   A comparison of PBX processor system characteristics.

busy hour. The busy hour calling rate is defined as the number of calls per hour that are presented to the system for processing during the busiest hour of the busiest day of a normal business week. It is measured in *busy hour call attempts* (BHCA).

The definition of busy hour call attempts is only concerned with the number of calls and has nothing to do with how long each call lasts. The reason for this is simple. Processor systems handle calls very quickly, usually within 20 or 30 milliseconds. While the majority of processing takes place during call setup, the processor system performs very little processing during a conversation, except for periodically checking to see whether the call has ended. Therefore, the amount of work the processor must perform depends only on the number of calls and has very little to do with how long those calls last.

Ideally, the busy hour call handling capacity of a processor system should increase proportionally with the number of users. If processor capacity does not increase as the system grows, there are two possible results. First, the system may be overpowered in small configurations and, therefore, more expensive than it should be. Second, the system may be underpowered in large configurations and, therefore, susceptible to overload during busy hours.

Another desirable feature in a processor system is the ability to adjust its capacity to fit user needs. One customer may have a low calling rate and should not have to pay for more capacity than needed. Another may have a very high calling rate and should be able to configure the system with a more powerful processor system for a reasonable additional cost.

In System A, the system-level processor must be provided in all system configurations. It is available in only one version with a fixed capacity. The module-level processors are added as each module is added. This processor system has excess power and is expensive in small line sizes. There is no opportunity to adjust the overall processor system power. However, the processing power does grow as modules are added. Given this configuration, the processor system should be rated at a certain number of BHCAs for a single module, with an incremental value added for each additional module.

The system-level processors in System B operate in a load-sharing manner. Any processor that is available can handle a request to process a call. This means that only a portion of the system-level processor capacity must be initially purchased. Processing power also increases as modules and ports are added. This incremental increase in processing power is closer to the ideal than System A's processor system. In addition, by adding or subtracting system-level processors, the overall power of the processor system may be adjusted to meet user needs. For example, in a situation with relatively few calls during the busy hour, such as a hotel or motel, only one or two system-level processors may be needed. On the other hand, in an ACD situation with a very high calling rate, eight or ten system-level processors might be used for the same-sized system. Given this configuration, the processor system should be rated at different levels depending on the number of system-level processors, plus an increment for each additional module.

Since System C has no system-level processor, the BHCA rating should be provided on a per-module basis and should grow as the number of modules increases. This should approximate the ideal growth curve. However, each module-level processor has a fixed capacity. Therefore, capacity adjustment is not easily accomplished.

Ability to remote parts of the processor system.   Although the switching network has the most significant impact on a PBX's ability to remote modules, processor system connections may also limit remoting capabilities.

In System A, remoting modules is straightforward. It is simply a matter of extending the individual connections from the system-level processor to the module-level processors. However, since this is a high-speed link (several megabits per second), fiber optics are required. The distance is limited to 25,000 feet and either a right-of-way must be available to run a private link or public fiber optic links must be available for lease.

In System B, the busses cannot be extended. Therefore, no part of this system can be remoted and all parts of the processor system must reside in the same equipment room.

In System C, module remoting capabilities depend on the characteristics of the LAN. Many PBX LANs have distance limitations between modules and/or the total distance across the LAN. If the module-level processor were connected with standard packet-network connections instead of a LAN, then modules could be remoted wherever packet-network connections were available.

Reliability.   The third criterion is reliability. PBX reliability may be evaluated in many ways, including mean time between failures (MTBF), mean time to repair, downtime, and availability. The parameter used most often used to describe PBX reliability is MTBF. It is formally defined as the total operating time during an interval, divided by the number of failures during the same interval. A more helpful interpretation of the MTBF is the average operating time between two failures. In addition, it is useful to distinguish between failures of the entire PBX system and failures of only a part of the system. The following discussion describes the parts of the PBX processor system that may cause a failure of either the entire system or a major part of it.

In PBXs that contain only one system-level processor that handles call processing, the consequences of a processor failure are catastrophic—no calls can be processed. As a result, almost all PBXs are provided with some alternative, either mandatory or optional, to avoid this situation.

There are three typical approaches for improving processor system reliability. One approach is to duplicate the system-level or the module-level processors. One of the processors processes calls, while the other is in a standby mode, ready to take over if the first one fails. This approach is expensive, but usually fairly effective with proper design.

A second approach is to use N + 1 redundancy, where N is the number of processors required to provide adequate processing capacity for the system. Rather than performing all call processing on one large processor, the PBX uses several smaller ones in a load-sharing configuration. The system includes one additional processor than is required to handle the call processing load. If one of the processors fails, the others handle the load until the failed processor is fixed or replaced. For a single failure, this approach is not much different than the duplication method because both systems continue to function at full capacity. However, if a second processor fails, the N + 1 system continues to run at reduced capacity, whereas the duplicated system fails. On the other hand, the N + 1 scheme may be more expensive. This method of improving reliability is usually applied only at the system level. The module-level and port-level processors are normally too small to be constructed from several smaller processors.

A third approach to processor reliability is to design the system so that each processor is responsible for only a part of the system. This approach is used in systems that have only module-level processors. Each processor is responsible for only one module. This may actually result in more points of failure than the duplicated system-level processor approach. Consequently, this configuration may not be more reliable overall. However, if a processor does fail, only one module is affected. In addition, the module-level processors may be duplicated in some systems.

In System A, the reliability of the processor system for total failures depends solely on the system-level processor. The module-level processors cannot contribute to improving this reliability because they do not operate independently. Since the system-level processor is such a critical component, a duplication option is offered which significantly increases the MTBF.

If a module-level processor fails in System A, only that module is out of service. The failure of a link between a module-level processor and the system-level processor has exactly the same effect. In this case, the percentage of the system that fails equals the percentage one module constitutes in the total system. That is, if a module-level processor or its link fails in a system with three modules, then it constitutes a partial failure of one-third of the system.

Since the module-level processors in System B cannot operate independently, the probability of system failures again depends on the system-level processor complex only. However, this system employs N + 1 redundancy. Therefore, if the number of processors has been properly chosen, full capacity is maintained during a single processor failure. If a second processor fails before the first is repaired, the system will operate on reduced capacity until the repairs are made. The failure of a module-level processor has the same effect as in System A.

The failure of a module-level processor in System C has the same effect as a module-level processor failure in either System A or System B—one module fails. However, since this may constitute a large part of the system, particularly in small PBXs, the module-level processors are duplicated.

Cost. In System A, the system-level processor complex contributes to the initial cost of the system. As each module is added, the module-level processors also add to the cost. Due to the system-level processor complex, the per-line cost of a small system (a few hundred lines) is relatively higher than Systems B and C. However, the cost of larger systems (a few thousand lines) is relatively lower.

In the System B load-sharing processor arrangement, it is not necessary to purchase all of the processors at once. Therefore, on a per-line basis, the initial cost should be lower than System A.

There is no initial cost due to system-level processors in System C. Thus, this type of arrangement has the potential to be economical at lower line sizes than either System A or B. However, if the modules are very large, then much of this potential advantage is lost. On the other hand, each of the module-level processors is part of a processor complex which is duplicated for reliability purposes. Thus, in large systems, the processor system becomes more expensive than in either System A or B.

Network Interface Circuits

What network interface circuits do

Network interface circuits allow various kinds of devices, such as telephones, data terminals, personal computers, and hosts, to be connected to the switching network. They also connect the switching network to facilities, such as trunks from other PBXs or central offices. In addition, some network interface circuits provide services such as tone supplies and conference bridges.

Network interface circuits perform three primary functions, including:

- changing the form of the signals from analog to digital, or from digital to analog
- multiplexing and demultiplexing
- providing the signaling used in sending and receiving control information

Network interface circuit overview

Since an interface circuit is required for each terminal or facility connected to the network, there are large numbers of these circuits in a PBX system. As a result, they have a significant impact on the overall cost and size of the system.

Network interface circuits are implemented in the form of circuit packs that plug into a circuit-pack carrier or shelf within a PBX cabinet. This arrangement is shown in Fig. 3.20a. The power converters, which are also included in this figure, are discussed in a later section.

When the circuit packs are plugged into the shelf, they make connections to one or more busses as shown in Fig. 3.20b. One of these busses is associated with the switching network function, while another is a control bus that allows the interface circuit packs to communicate with the processor system. (Alternatively, the function of the control bus may be accomplished by using some of the switching network paths.) Finally, there is also a power bus that supplies power to the circuit packs.

Logically, the wires connecting the interface circuit packs to various trunks and terminals appear as shown in Fig. 3.20b. The signals enter the wires on one side of the circuit pack, are processed, and then leave the other side of the circuit pack. Physically, the wires to the trunks and terminals are usually connected to the back of the circuit pack.

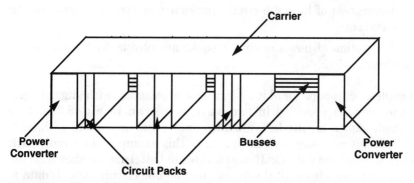

Figure 3.20a   Network interface circuit pack carrier arrangement.

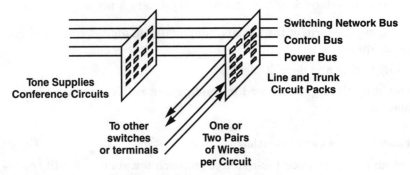

Figure 3.20b   Logical network interface circuit electrical connections.

There are a number of different types of network interface circuit packs including those that handle analog lines, digital lines, analog trunks of several kinds, digital trunks, and service circuits.

One other important interface circuit is the one used to connect the attendant console or consoles to the switching network. Attendant consoles are used by the PBX attendants to handle incoming calls as well as other calls that require special handling. In the past, these consoles had direct connections to the trunk circuits or other special connections into the system. However, today most attendant consoles are of the *switched loop* variety. That is, the consoles have a number of loops, which are similar to the lines connecting any other terminal to the PBX. These loops are used to connect calls to the console. Even though the consoles are connected to the switching network in the same way as digital lines or trunks, the PBX identifies them as attendant consoles and gives them priority treatment.

The model of the network interface circuits used in this book includes:

- the types and numbers of circuits included on each circuit pack

- a description of how the circuit packs are electrically connected to the system

- a description of how the circuit packs are physically housed in the system

Examples. Support for data terminals associated with digital telephones can be provided in two ways, as shown in Fig. 3.21a. One method, shown on the left, provides data capability along with each voice circuit on a single circuit pack. In this example, the circuit pack handles voice and data for 16 users. A second alternative, shown on the right, only provides digital voice circuits on one circuit pack. If data is required, another circuit pack must be added. In this example, the first

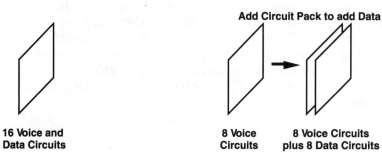

Add Circuit Pack to add Data

**16 Voice and
Data Circuits**

**8 Voice
Circuits**

**8 Voice Circuits
plus 8 Data Circuits**

Figure 3.21a Two methods of providing for data.

circuit pack provides voice capability for eight users. Adding the second circuit pack provides data capability for those same eight users.

Figure 3.21b shows two methods of accommodating T-1 digital trunk circuits. In the universal slot design, shown on the left, each slot in the circuit pack carrier will accept *any* network interface circuit pack. Therefore, the wiring, which extends the circuit connections outside the cabinet, must accommodate the maximum number of wire pairs that *any* circuit pack might require. In this example, 24 pairs of wires are provided for each circuit pack carrier slot. The T-1 digital trunk circuit shown uses a single carrier slot.

An alternate approach to the universal slot approach is the motherboard design shown on the right in Fig. 3.21b. Wiring across the bottom half of each carrier slot is used to distribute the switching network, control, and power busses. The top half remains unused until the types of circuits that are required have been determined. A motherboard, unique to the circuit packs being installed, is then added to provide the proper number of wires needed to extend the circuits outside of the cabinet. This ensures that the wiring to each slot is exactly what is needed. In this example, two pairs of wires are needed to accommodate a T-1 digital trunk circuit. The trunk circuit itself is provided on 10 circuit packs and requires 11 circuit-pack carrier slot positions.

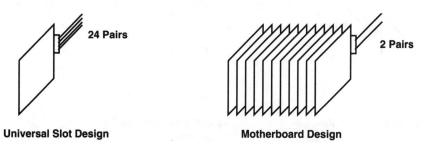

24 Pairs

2 Pairs

Universal Slot Design

Motherboard Design

Figure 3.21b Two methods of accommodating T-1 digital trunk circuits.

Finally, Fig. 3.21c shows two approaches to installing digital line circuit packs in a universally slotted carrier. The circuit pack on the left provides circuits for 16 users with voice and data needs. In this example, only one pair of wires per user is required and a total of 16 pairs are provided to each universal slot. The circuit pack on the right provides circuits for eight users with voice and data needs. In this example, the circuit pack shown requires two pairs of wires per user. However, a total of 24 pairs of are provided for each slot in the carrier.

Criteria for evaluating network interface circuits

Five of the most important criteria for evaluating network interface circuits are:

- density of the circuit packs

- cost

- ability to use space efficiently

- ability to use the wiring efficiently

- reliability

Density of the circuit packs.   The interface circuits, particularly the line and trunk circuits, along with the carriers that contain them, constitute the largest portion of the hardware in most PBX systems. As a result, they have a significant impact on the physical size of the system. Since most circuit packs and circuit pack carriers are roughly the same physical size, the critical factor is the number of circuits that can be accommodated in a given space. This parameter is measured in terms of density. The *density* of a group of circuit packs is

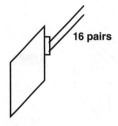

16 pairs

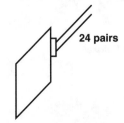

24 pairs

16 digital line circuits
(1 pair)

8 digital line circuits
(2 pair)

Figure 3.21c   Two approaches to digital line circuits in universal carriers.

defined as the number of circuits supported by the group divided by the number of circuit pack slots required to house the group. For example, in Fig. 3.21a the circuit pack on the left provides 16 voice/data circuits and requires one slot. The density is 16 voice/data circuits per slot. On the other hand, the circuit packs shown on the right in Fig. 3.21a provide eight voice/data circuits on two circuit packs that require two slots. The density is four voice/data circuits per slot. The density of the 16-circuit packs is four times that of the alternate approach.

The T-1 digital trunk circuits shown in Fig. 3.21b provide a second example. The trunk circuit pack on the left provides one T-1 trunk circuit and takes one slot. The density is one T-1 circuit per slot. The group of circuit packs on the right also provide one circuit, but they require 11 slots. The density is one-eleventh of a T-1 circuit per slot.

As a final example, assume that a group of three circuit packs is required to provide circuitry for 16 analog trunks. However, due to the physical dimensions of the components on two of the circuit packs, two slots are required to accommodate each one. Therefore, a total of five slots are needed for 16 trunk circuits. The density is 3.2 trunk circuits per slot.

Cost. Since the circuit pack carriers and the interface circuits they contain constitute the largest portion of the hardware in most PBX systems, they, along with the terminals and the wire, are the most critical elements influencing the per-line cost of the system. In general, for interface circuits, the cost per circuit will be lower for circuit packs with higher densities. Therefore, a T-1 interface that requires only one circuit pack should be much less expensive than the one that requires ten. A line circuit pack with a density of 16 voice/data circuits per slot should be less expensive than the packs that support only four voice/data circuits per slot.

Ability to use space efficiently. The motherboard design shown in Fig. 3.21b requires that a motherboard be installed before installing the associated interface circuit packs. Motherboards cover several slot positions. The actual number depends on the particular motherboard. If all the slot positions are not used after the motherboard has been installed, the remaining slots may only be used for interface circuits of the same kind. Therefore, it is possible to have empty slots that are not usable because they already have the "wrong" types of motherboards installed. In addition, it is possible to have empty space in a circuit pack carrier that cannot be used because a particular motherboard is too large for the available space. On the other hand, in the universal slot design, any circuit pack may be put in any slot.

Ability to use wiring efficiently. The universal slot design requires that each slot provide for the maximum number of wire pairs that will be used by any circuit pack. Often many of the available wires are not used. For example, the T-1 trunk interface in the universal slot design shown in Fig. 3.21b uses only two wire pairs even though 24 pairs are provided. The universal slot design may also limit a manufacturer's ability to increase the density of the circuit packs. For example, in Fig. 3.21c, the circuit pack that provides eight circuits has 24 pairs of wires available to it. Suppose however that, due to improved technology, it is possible to put 16 circuits on the same size circuit pack. There are not enough wires in the universal slot to support it. Thirty-two pairs of wires are required and only 24 are available. If a new universal slot carrier is designed to accommodate the circuit packs with 16 circuits, the new circuit packs will not fit in the old carriers. This is not a problem with the motherboard approach. A new motherboard that is designed to accommodate the new circuit packs will fit into the old carriers.

Reliability. Reliability is not as important for the network interface circuits as it is for the switching network or the processor system. This is because the failure of a network interface circuit affects only a few lines, trunks, or service circuits. These failures are not considered as serious as a failure of the whole system or a large part of it. On the other hand, the large numbers of interface circuits may be the largest source of maintenance activity if they fail frequently.

Terminals

What terminals do

Terminals are the devices that provide the users with access to the system. They include analog and digital telephones, data terminals, and PCs. Terminals provide access to other terminal users, computers, or other services by using switched connections provided by the system. Terminals must provide a convenient means for the user to request service from and control the system by pushing buttons, dialing, typing on a keyboard, or selecting an item from a menu. Terminals adapt the input information from the user to a form that is suitable to be sent to the system. For example, they convert voice to the appropriate electrical signals or multiplex the information from a data terminal into a digital information frame that is transmitted to the system. Terminals also adapt the information received from the system to a form that is compatible with the user's needs. For example, they convert electrical signals to sound waves or display information on a screen.

Terminal overview

Since most users' only interaction with the system is through a terminal, there are a number of important aesthetic and ergonomic issues associated with them. However, to provide a suitable architectural model, the most important characteristics of a terminal are:

- the functions it performs
- how it is connected to the system
- how it interconnects with other terminal products on the desktop

The simplest terminal is an analog telephone. It typically provides:

- a transmitter/receiver device called a handset
- a rotary or pushbutton dial
- an alerting device, which is usually a ringer

An analog telephone is connected to the system with a single pair of wires. It communicates to the system with either electrical current flow changes or a series of tones consisting of predefined frequencies.

Digital telephones have significantly more capabilities than their analog counterparts. A block diagram of a typical digital telephone is shown in Fig. 3.22a. The main components include:

- a codec
- a microprocessor
- a multiplexer

Digital telephones that support data terminals as well as voice conversations also include a data module.

The *codec* samples and digitizes the analog voice signal from the transmitter and delivers the digitized output to the multiplexer. This digital information channel is called the Information 1 (I1) channel. The codec also receives incoming digitized voice from the multiplexer on the I1 channel. It decodes the digital information and supplies the resulting analog signal to the receiver.

The microprocessor provides the intelligence for the telephone. Through an I/O interface, it receives information from the buttons and the dial. It also sends control information to the lamps, display, and alerting functions through the I/O interface. A memory associated with the microprocessor stores information required to provide such features as a clock, calendar, memory dialing lists, automatic redial information, and other features. The microprocessor processes the inputs from the buttons and the dial pad and sends a stream of digital control information to the multiplexer via the signaling (S) channel. The microprocessor also receives control information from the multiplexer on the S channel. This information is processed and then used to control the display, lamps, and alerting function.

The multiplexer receives the digital information from the codec and the microprocessor and organizes it into an information frame to be sent to the system via the transmit circuit. The multiplexer receives information from the system via the receive circuit. After demultiplexing this information, the appropriate parts are sent to the codec and microprocessor via the I1 and S channels, respectively.

As previously discussed, if a data terminal is to be supported, the digital telephone includes a data module. Physically, this data module sits under the telephone, is attached to the side of the telephone, or is provided on a circuit pack which is installed inside the telephone. Typically, as shown in Fig. 3.22a, it provides an RS-232C connector for attaching the data terminal. Other data connections may be supported with appropriate data modules. Electrically, the data module converts the data from the terminal into a format suitable for multiplexing with the voice and signaling information. The converted information is sent to the multiplexer on the Information 2 (I2) channel. Information is also received from the multiplexer on the I2 channel. The data module converts this information into a form suitable for use by the terminal.

A typical proprietary information frame that is created by the multiplexer and subsequently transmitted to the system is shown in Fig. 3.22b. A similar frame is received from the system. The S, I1, and I2

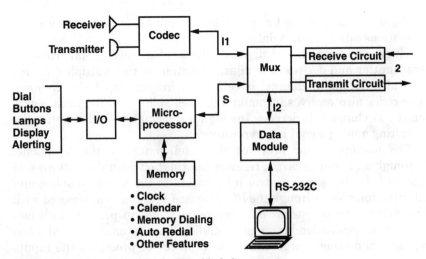

Figure 3.22a A typical digital telephone block diagram.

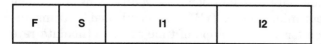

Figure 3.22b   Proprietary information frame.

fields which were described above are shown. The additional field, called a framing field (F), is a number of bits used by the system to identify the beginning of the frame.

Usually, a proprietary information frame, such as the one just described, is repeated at the voice sampling rate of 8000 times per second. In most cases, the I channels include eight bits per frame and are, therefore, 64-Kbps channels. The S channel and the number of F bits vary from one manufacturer to another. However, typically eight bits are used for each of these functions, yielding a 64-Kbps S channel and 64 Kbps for framing. In this case, the overall bit rate for the entire information frame is 256 Kbps. It's important to note that any digital telephone that implements this type of proprietary information frame will usually only work with PBXs from the same vendor.

A new standard information frame format, called the Integrated Services Digital Network (ISDN) Basic Rate Interface (BRI), has been recommended by the CCITT. One of the objectives of this standard is to allow the interconnection of any ISDN BRI digital telephone to any ISDN PBX. Figure 3.22c shows a simplified representation of an ISDN BRI information frame for PBX systems. This frame repeats at 4000 times per second. It contains several framing (F) bits, a 16-Kbps signaling channel (D), a 16-Kbps control channel for a service called "passive bus" (D_e), and two information (B) channels at 64 Kbps. The overall frame data rate is 192 Kbps. The ISDN BRI specification calls for two pairs of wire, one for transmitting and one for receiving.

Examples of terminals

Figure 3.23 provides some examples of different terminal architectures. Many manufacturers' digital telephone sets use two pairs of wires as shown in Fig. 3.23a. Other manufacturers use only one pair of wires as shown in Fig. 3.23b. This is done by using either balanced hybrids or time compression multiplexing (TCM). *Balanced hybrids* are specially wound transformers that, when properly matched to the

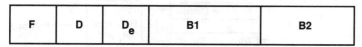

Figure 3.22c   ISDN BRI information frame.

line impedance, allow data transmission in both directions on a single pair of wires without interference. *TCM* is accomplished by transmitting in one direction for a short period of time at twice the data rate used inside the telephone or PBX system. The direction of transmission is then automatically reversed and information is transmitted in the other direction for an equal period of time, also at twice the internal data rate. If this change of direction is repeated at frequent regular intervals, the resulting information flow appears as if it were accomplished by transmitting at the internal data rate on two wire pairs.

The capability of the terminal to communicate with the system is determined by the structure of the information frame. For example, suppose an information frame is transmitted 8000 times per second on a single pair of wires using TCM. The signaling channel consists of one bit at a data rate of 8 Kbps. The voice channel consists of eight bits at the standard 64-Kbps data rate. However, the data channel consists of only two bits; thus, the maximum data rate for this channel is 16 Kbps. This is a fundamental architectural limitation when compared to other systems that transmit up to 64 Kbps of data and voice simultaneously.

Another way in which PBX architectures vary is the method used for connecting PCs and digital telephones at the same location to the system. In Fig. 3.24, Arrangement A shows a typical installation. A circuit pack is inserted into a spare slot in the PC. The pack is connected to a digital telephone and the telephone in turn is connected to the wires going to the system. In this case, a proprietary link is used between the telephone and the system.

Arrangement B is a slight variation of Arrangement A. In Arrangement B, the telephone set connects to the card in the PC, and the card connects to the wires that go to the system over a proprietary link. Arrangement C is much like Arrangement B, except that it uses a less expensive analog telephone and provides a standard ISDN BRI inter-

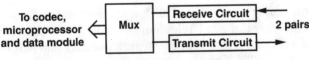

Figure 3.23a   Digital telephone—Two-pair arrangement.

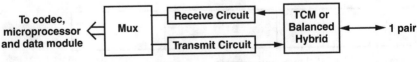

Figure 3.23b   Digital telephone—One-pair arrangement.

face to the system. Finally, Arrangement D shows a case where a special digital telephone is used. The telephone has an RS-232 connector which plugs into an asynchronous port on the PC.

Criteria for evaluating terminals

The five most important criteria used in evaluating terminals are:

- cost
- number of wire pairs
- capabilities allowed by the frame format
- interaction of digital telephones with personal computers
- the data terminals supported

Cost. Terminals, along with the network interface circuits and the wire, are the most critical elements influencing the per-line cost of the system. Analog telephones are less expensive than digital telephones. However, activating features from an analog telephone requires dial-

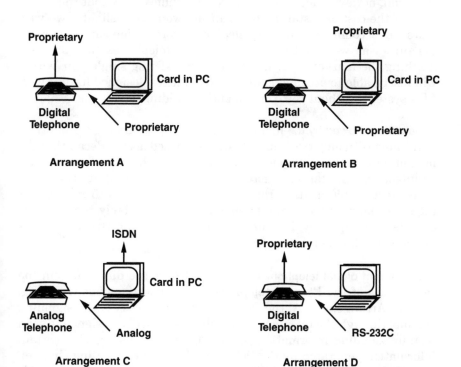

Figure 3.24 Typical implementations of personal computer and telephone interconnections.

ing feature codes, usually an asterisk (*) or a pound sign (#) followed by one or two digits. These codes are hard to remember unless they are used frequently. With digital telephones, on the other hand, features are activated by pushing buttons which are labeled with the names of the features.

Since digital telephones are more expensive than analog telephones, many system managers provide digital telephones only to those users with a perceived need for high-functionality instruments. Other users are provided with analog telephones. This approach, however, necessitates stocking both kinds of telephones and interface circuit packs in inventory. Also, when a user moves from one location to another, the telephone must match the interface circuit serving the new location. One solution to this problem is to trade the higher cost of providing all users with digital telephones for a possible savings in inventory and change activity.

Number of wire pairs. If digital telephones use only one wire pair, it may be possible to install them using single-pair wiring already in place. This is a potential advantage in existing buildings. In new construction, however, detailed cost analysis studies show that the difference in the cost of installing one pair or two is very slight. Two wire pairs will accommodate digital telephones from a wide variety of manufacturers as well as the new ISDN BRI telephones that are now reaching the marketplace. Locations with existing single-pair wiring will not be able to accommodate BRI telephones designed for use with PBX systems without rewiring an entire building or campus.

Capabilities allowed by the frame format. A proprietary digital information frame with only two data bits was described above. Recall that the maximum data rate supported was 16 Kbps. This is a fundamental architectural limitation. Increasing the data handling capability requires a change in the frame format. Changing the frame format requires changing the digital telephones and, most likely, the interface circuits. Consequently, it is important to understand the capabilities provided by the information frame format.

Interaction of digital telephones with PCs. There are differences in the capabilities of the PC and telephone interconnections shown in Fig. 3.24. In Arrangement A, the PC can only gain access to the signaling channel through the telephone. The telephone may be designed to send certain signaling information to the PC. It may also receive certain information generated by the PC and put it in the signaling channel. Ultimately, however, the ability of the PC to provide control depends upon the design of the telephone. In Arrangement B, on the other hand, the PC has direct access to the signaling channel. This allows the PC to provide more control over various calling situations.

To illustrate the difference in the two approaches, suppose that an incoming call control feature is desired. The feature should allow the user to enter a list of phone numbers into the PC. Any incoming call from one of these numbers should ring the telephone. All other calls should be redirected to a voice mail system. This is easily accomplished using Arrangement B. The PC receives, over the signaling channel, the calling party identification along with an alerting command. If the caller's number is on the list, the PC sends an alerting command to the telephone. If the caller's number is not on the list, the PC instead sends a forwarding command to the system, which in turn forwards the call to the voice mail system. In contrast, when the telephone in Arrangement A receives an alerting command, it may ring the telephone and not pass the alerting command or the caller identification to the PC.

At first glance, Arrangement C seems to provide an advantage because inexpensive analog telephones are used in place of more expensive digital telephones. However, an analog telephone is quite limited in function, requiring that most features be activated from the PC. If the PC does not provide multitasking operation, that is, the ability to process more than one task at a time, handling telephone calls may interrupt work being performed on the PC.

Arrangement D has the advantage that no special circuit pack is needed in the PC. Another advantage of this arrangement is that the RS-232 connection accepts standard modem commands. Therefore, any of the commercially available software communication packages may be used. The other three arrangements only work with software supplied by the vendor who provides the telephone-to-PC interface circuit pack that plugs into the PC.

The Power System

The method used for powering a PBX may have a significant impact on the cost and reliability of the system. Typically, in a modern PBX, a power loss results in a complete failure.

What it does

The power system for a PBX supplies the power necessary to run all parts of the system including the telephones.

Power system overview

Ultimately, the power is derived from an alternating current (AC) source. However, the circuitry inside a PBX uses direct current (DC). Therefore,

AC power must be converted to DC power. This is done by a *rectifier.* Furthermore, the DC voltage level from the rectifier must be changed into a range of DC voltage levels that are required by the various parts of the system. This is done with *DC-to-DC converters.* Typical voltages supplied by these converters are: +12 volts, –12 volts, +5 volts, and –5 volts.

The main power source to the system may be supplied in one of two ways. In the first method, the system requires DC input as shown in Fig. 3.25a. In this case, the rectifier is external to the system. It is not usually priced as part of the system and it takes up additional floorspace. If battery backup is provided, the batteries are installed between the system and the rectifier.

In the second method, the system is powered with AC as shown in Fig. 3.25b. In this case, the rectifier is inside the system. However, if battery backup is required, it may require additional components external to the PBX. The usual method of providing battery backup for an AC system is to install an external rectifier to charge the batteries. The battery voltage must be changed back to AC before entering the system. This is done by a power device called an *inverter.*

Inside of either type of system, the power distributed within the cabinets is usually DC. This may be in the form of multiple voltages from the main rectifier, or a single voltage, usually –48 volts DC. A single-voltage distribution system requires additional DC-to-DC converters to accommodate the different voltage levels required by various components. These converters are usually located near the circuits that will use the power. One possibility is to locate the converters behind the circuit pack carriers. Another possibility is to locate them in the circuit pack carriers or shelves. The converters are often provided as pairs in a load-sharing arrangement. If one converter fails, the other provides the required power.

Criteria for evaluating the power system

The important criteria for evaluating the power system are:

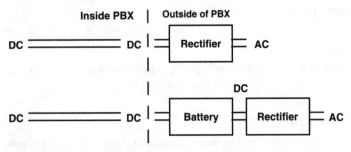

Figure 3.25a External AC-to-DC power conversion.

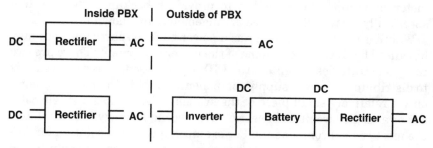

Figure 3.25b   Internal AC-to-DC power conversion.

- reliability
- cost—both hardware and operation
- AC or DC supply to the system

Reliability. When commercial power fails, the effect on the PBX depends on the power backup strategy. There are three typical cases:

- no power backup
- battery backup for memory only
- full system power backup

If no power backup is provided, the entire system is out of service during a failure of the main power source. However, if there are analog trunks connected to the system, a power failure emergency transfer system may be provided to connect a few analog telephones to the analog trunks. When power is restored, the system will reload the memory from the memory backup system. This may take anywhere from a few minutes to a half hour, depending on the size of the system and the backup mechanism employed.

In the second alternative, batteries are provided to back up a portion of the memory for a short period of time, usually a few minutes. This does not prevent the system from going out of service, but does shorten the time required to restart the system when power is restored. This is because the memory does not have to reloaded. However, if the power failure lasts longer than the time that the batteries are capable of sustaining the memory, the system will have to reload the memory when power is restored.

In the third alternative, backup power is provided for the entire system. This alternative is only practical if operation during commercial power outages is important. There are two ways to provide full system power backup. The first is to provide some type of uninterruptable AC source either with a power system that has backup batteries or a motor-generator. The second is to provide a backup battery system specifically for the PBX. These arrangements are shown in Fig. 3.25.

Whether or not the system has a power backup source, outages may be caused by the failure of some of the power equipment. There are two common strategies to protect the PBX from these failures. The first is to distribute the power supply components so that if a problem occurs, only a relatively small piece of the system fails, such as a single cabinet or a single circuit pack carrier. The second strategy is to duplicate the most critical components. For example, a common solution for AC-powered systems is to provide one rectifier per cabinet, distribute −48 volts to the circuit pack carriers, and duplicate the DC-to-DC converters associated with each circuit pack carrier.

Cost.   Unfortunately, power systems that come is large units cost less than those that use a number of small units to provide the same capacity. This goes counter to the strategy of distributing the power system for reliability reasons. Many of the tradeoffs related to the power system involve balancing cost against reliability. In addition, the power required to operate the system is often a significant expense. Some digital telephones operate on as little as one-half watt of power. Others require three or four times as much. Overall, the power requirements for different PBX implementations may vary greatly. One system may require twice as much power as another system of the same size.

AC or DC supply to the system.   Determining whether the AC or DC approach is better for a particular situation depends heavily upon the system power backup plan. If no backup is to be provided or if an uninterruptable AC source will be used, a system designed for AC input is probably the least complex and expensive. On the other hand, if a battery backup system is planned to support the PBX, a system designed for DC input will require fewer external parts and is probably the least expensive alternative.

Future Trends in PBX Architecture

In previous sections of this chapter, you learned how to analyze and compare current PBX architectures. In this section, you will learn about trends in the telecommunications industry that will cause significant changes in PBX architectures. Some of these trends may have an impact as profound as the invention of dial telephones, stored program control, or digital time-division switching. The five which seem to have the most potential for creating these changes are:

- the incorporation of ISDN technology
- the increased computing power in terminal products

- the increasing use of host computers to control PBX applications
- the increased reliance on highly distributed architectures
- the integration of voice and data

Integrated Services Digital Network

The emerging worldwide implementation of ISDN has been characterized by Theodore Irmer, director of the CCITT, as the most significant communications event since the invention of the dial telephone. This technology will certainly have a profound effect on PBX system architecture.

Definition. *ISDN* is a shared network that provides end-to-end digital transmission and supports a wide range of services. Access to this network is limited to a small set of well-defined *standard* interfaces. These interface standards define a rich set of messages that may be used to communicate with and control the network.

Examples. From a PBX perspective, one of the goals of ISDN is to standardize the PBX-to-digital telephone interface. In the near future, this standard interface will allow any ISDN telephone to be connected to any ISDN PBX as easily as an analog telephone is connected to any PBX today. This means that, instead of being forced to buy digital telephones from their PBX vendor, users will be free to purchase digital telephones from any ISDN telephone vendor. Not only will users have a much wider selection from which to choose, but competition should drive down the cost of digital telephones.

Furthermore, as the difference in price between analog telephones and digital ISDN telephones narrows, it will become practical to provision a PBX system with digital telephones only. This will allow users to stock only digital interface circuit packs rather than both analog and digital circuit packs. It will also reduce the number of different kinds of telephones that must be stocked. Also, moving telephones will be easier because the wires at each work location will always terminate on the correct type of interface circuit pack.

ISDN will also have an impact on PBX-to-trunk connections. Traditional trunks connecting PBXs to the public switched network are dedicated to a particular service (for example, WATS) because this is the method by which the network determines the service being requested. ISDN trunks do not need to be dedicated to a particular service because they are accompanied by a signaling channel which may be used by the PBX to tell the network what service is being requested. Thus, ISDN trunks may be installed in one large group rather than several small groups which are dedicated to a particular service. Since large trunk groups are more efficient than small trunk groups, ISDN trunks potentially provide a substantial cost savings.

In the future, PBXs will be designed to support ISDN interfaces only. The architecture of these systems will be significantly simplified by not having to accommodate analog telephones and several different kinds of analog trunks. Consequently, these systems will be less expensive than today's typical PBX.

Smart terminals

The price of computer and memory chips is steadily decreasing. This makes it practical to economically increase the processing power of digital telephones. In addition, the increased use of PCs makes it possible for many telephones to take advantage of their processing power.

Definition. *Smart terminals* are either digital telephones with a processor and memory or telephones connected to a PC. The processing power of smart terminals, along with the signaling channel in the digital link to the PBX, provides these terminals with an opportunity to participate more fully in the operation of the PBX.

Examples. Most of the digital telephones in use today communicate with the PBX processor on a *stimulus* basis. This means that when a button is pushed on a telephone, the telephone simply reports that the button was pushed. For example, if button three was pushed, it is up to the system to examine the translation information in memory to determine which service (such as call forwarding or call pickup) is associated with that button on that particular telephone. Storing this information for each telephone requires significant amounts of memory, and searching for this information on each call requires significant amounts of processing time. Furthermore, the system administrator must spend a significant amount of time initially loading this information into the system as well as making changes each time telephone features are changed.

With smart terminals, communication with the system may be *functionally* oriented. That is, when a button is pushed on the telephone or a command is typed on the keyboard, a specific command is transmitted, such as, "make a voice connection to the voice mail system." The system does not need to know how the message was generated. Therefore, the terminal user is free to generate the message in any way that is convenient. These messages could be generated by pushing a button or by selecting from a menu on a PC screen. The operation may be customized to suit the needs of the user. Furthermore, any desired operational changes may be made by the user instead of the system administrator. This eliminates the necessity of telling the system administrator what is wanted and then waiting for the change to be made.

Also, with a functionally oriented smart terminal, the PBX immediately knows what is being requested when the message is received. Therefore, it does not have to waste valuable call processing resources searching for the meaning of a particular button push. Thus, in the future, much of the work that has been done by the PBX itself or by a system administrator will be performed by the terminal user or automatically by the terminal.

Host computer control

Host computers have been used to collect, store, and process information from PBX systems for a number of years. For example, call detail information from a PBX is typically passed to a host computer for generating reports that are used to distribute phone bills to the appropriate departments. Now host computers are being used to control the PBX. As a standard interface for this connection emerges, it will become common for a host computer to make many of the high-level decisions that are currently made by PBX processors.

Definition. Host computers connected to a PBX processor allow the host to control the PBX as well as receive information from it. This allows a user to customize the operation of the system by changing a program on the host computer.

Examples. Host computers in conjunction with PBXs have many telemarketing applications. For example, incoming calls to a PBX from an "800" sales line are distributed to a group of customer service agents. Using a connection from the PBX processor to a host computer, the PBX passes the identification of the caller to the computer along with the number the caller dialed and the identity of the agent to which the call will be assigned. The host computer then displays on the selected agent's terminal a screen of information about the caller. This all takes place before the call is answered. This example involves one-way information transfer from the PBX to the host.

Another example involves outgoing call management. In this example, the host computer is programmed to select a telephone number from a list of sales prospects. The host then instructs the PBX to dial the number. If the prospect answers, the call is switched to an agent, and the screen on the agent's terminal is filled with information pertaining to the called party. If the prospect doesn't answer, the host instructs the PBX to disconnect the call and dial the next number on the list. This example involves two-way communication between the PBX processor and the host.

One final example illustrates a case where the host has almost total control over the PBX. In cellular mobile telephone systems, there is a need to monitor the signal from the mobile unit and to connect a channel from the antenna with the best signal to the public switched telephone network. This is done by a computer that monitors signal strengths and makes decisions about which antenna to use. The computer then causes a switch—a PBX in many cases—to make the appropriate connections.

Distributed architecture

PBX architectures are tending toward interconnection of smaller, more specialized units. For example:

- Some are physically composed of stacks of circuit pack carriers.
- Some have processing distributed to each module.
- Some have small switching units interconnected by standard digital trunk facilities.
- Some work with specialized units called servers which provide additional features, such as voice mail or FAX mail.

Definition. A PBX becomes *distributed* when many of the functions previously performed by the PBX are apportioned among a number of smaller, more specialized, functional units that are interconnected by standard interfaces.

Examples. This trend began with:

- physically modular units
- distribution of processors throughout a system
- small switching units interconnected with standard digital facilities

It is continuing with PCs taking over many of the functions that PBXs have previously provided for terminals, and host computers taking over many of the high-level call processing decisions. Ultimately, future business telephone systems may be constructed from a number of functionally specialized units, including:

- storage units, such as voice mail, FAX mail, and database servers
- switching units, including network components from PBXs and LANs with voice transmission capabilities
- terminal units, such as smart telephones and PCs
- control units, such as PCs and hosts

By combining existing products with only minor extensions of their current capabilities, a potential architecture might be as follows:

- The basic switching unit is a small TSI. It provides nonblocking connections for 20 to 200 users. It has a controller which keeps track of details and makes connections on command.

- The basic control unit is a PC. The switching unit controller is connected to the control unit through a small computer system interface (SCSI) port, which is a high-speed I/O interface. The control unit is connected to a host computer with a token-ring LAN.

- The terminals are ISDN BRI telephones, most of which are connected to PCs.

- The switching units are connected together with ISDN digital trunks. The control units communicate over the signaling channel of the ISDN trunk interface. Switching units can be located anywhere that ISDN trunks are available, creating a potentially worldwide system.

- Voice mail, text mail, and FAX mail units have their own PC control units. These mail units are added to the system by attaching their control units to the token-ring LAN and by making ISDN connections to the switching units as required.

The components for such an architecture already exist. By using a small number of existing standard interfaces, they can be combined into a flexible worldwide business communications system.

Voice/data integration

The appropriate role for the integration of voice and data in a PBX is a hotly debated subject. At one time, many people thought that the PBX would become the premier voice/data controller. Since then, LANs have significantly increased in popularity and only a small percentage of all PBX lines are used to switch data.

Definition.   There are three types of potential voice/data integration.

- *Physical integration*—the combining of voice and data units in one physical housing. These units are not necessarily connected electrically. Integrated Voice/Data Terminals (IVDTs) provide one example of physical integration.

- *Electrical integration*—multiplexing of voice and data signals on the same transmission path or electrically connecting voice and data units to provide communication between them. This could occur in a terminal, on the connections between a terminal and a PBX, in the PBX itself, or on transmission paths within a wide area network (WAN) or metropolitan area network (MAN).

- *Functional integration*—integration of applications as seen from the user's point of view. For example, the ability to attach voice comments to a text document would be an integrated application. Functional integration is probably the most important type of integration to a system user. It is usually accomplished by some form of electrical integration.

Examples.   Some of these integration possibilities may affect the architecture of a PBX. They will be discussed in the order mentioned above.

Physical integration.   IVDTs without electrical or functional integration can be used with a PBX, but do not seem to provide much benefit. They require separate wiring to the system and are essentially just a terminal and a telephone in the same plastic housing. LANs can be put in the PBX cabinet, but again the wiring is separate and there is little apparent benefit.

Electrical integration.   A number of methods of providing electrical integration or interconnecting telephones and PCs were discussed previously. If it is deemed desirable to switch data from a PC through a PBX, then electrical integration of the voice and data in either a digital telephone or a PC avoids the necessity of providing separate wiring for the two devices. There is electrical integration in the PBX in the sense that both voice and data signals are handled in the same way. Electrical integration on transmission facilities between switches is not new. However, some significant economic advantages are possible by doing so.

Functional integration.   Clearly, functional integration is already affecting PBX architectures. This chapter has previously mentioned two examples of functional integration, including:

- connections between digital telephones and PCs
- connections between the PBX processor system and a host

This section provides two more examples of functional integration from the user's point of view. In the first example, a user makes a new type of call to another user. The connection established by the system includes both voice and data channels that allow the called party to view, on his or her terminal screen, the same information that is on the caller's terminal screen. While they talk, both parties can make changes to the terminal screen information and see any changes that are made. This capability is called *screen sharing*. Two or more parties may be involved and the voice call, along with the screen information, can be transferred to a third party. In conference situations, the names of the participants can be displayed on the screen along with the identity of the current screen's originator.

The second example involves a connection in which one of the parties has a PC and the other party has a Group 4 (digital three-second transmission) FAX machine. Images can be sent from the PC to the FAX machine or from the FAX machine to the PC. During the conversation, images are transmitted back and forth as each person modifies the information.

4

Features and Applications Processors

Introduction

Features are all the services the PBX performs for the user community. Some are overt, such as Call Forwarding, requiring the user to request them. Some of them, such as Least Cost Routing, are performed behind the scenes without the user even realizing they are happening. Some of them, such as Automatic Trunk Testing, are to assist the telecommunications organization in managing the PBX. Features turn a raw telecommunications switch into a useful business tool. How useful the tool is depends on the particular business enterprise, what features are provided, how they are implemented, and how skillfully the user community can use them.

Applications processors are auxiliary processors that provide some of the features for the telecommunications system. They are connected to the PBX call processing processor via a communications channel so that they can share information, but the processing is entirely separate. They do not share common databases, though they may have a replica of a switch database such as the administration database. This chapter will also discuss why applications processors evolved, how they are used, what some of the tradeoffs are regarding their use, and where development will go.

Modern PBX Features

Evolution of PBX features

PBXs began as small switching systems on customer premises as a means of reducing the number of pairs required to connect the telephones to the local central office. The number of pairs were reduced because all telephones shared a few common pairs to the central office (trunks) and intraoffice calls did not switch through the central office.

Early PBXs were manual, so callers always talked to a live person. A person possesses a marvelous device weighing about three pounds called a brain. This brain operates some wonderful I/O devices called eyes, ears, and mouth. It also controls the body, commanding it to do such things as walk and talk. Put them all together and the most wonderful collection of features are available to the user because the attendant could do all sorts of things besides insert and remove plugs from a cord board! Thus were born PBX features. When the attendant was replaced with the automatic switch, the user community was incensed: "How can I find Mr. Burns in an emergency when he's not at his desk!" or "Have Bill call me as soon as he gets off the phone." These and similar problems were the motivation behind PBX feature development. Until about fifteen years age, most PBX feature development was really just attempts to make the automatic system do what Milly used to do for you when you called.

About fifteen years ago, venders started developing features in response to business problems that could be solved with the telecommunications system. Some of these were in response to the evolving long distance tariffs such as Automatic Alternate Routing and Automatic Route Selection. Some were in response to the need to control telecommunications costs such as Station Message Detailed Recording and Traffic. Some were in response to new communications needs such as data communications. And some were specialized applications for specific business sectors such as Hospitality features for hotels and motels.

Vendors have been developing PBX features for several decades in response to business needs. Consequently, there are several hundred available on most PBXs. This is a list that even those working with them regularly find hard to comprehend! To help facilitate understanding, in the following sections, PBX features will be discussed as *logical groups*. These are my logical groupings since no widely accepted standard exists and, as features continue to be developed, new categories will undoubtedly emerge.

- System features
- Station features
- Messaging features
- Networking features
- Data Communications features
- Call Center features

- Specialized Business features
- System Management features
- Maintenance features

Each of these feature groups will be discussed in the following paragraphs. Examples of features in each of these groups will be given, but complete lists will not be provided because different vendors give very similar features different names.

System features

System features are those features provided by the system to the system as a whole, but are not part of another feature group. For example, the attendant features are a common service to the entire user community but are not associated exclusively with any other group. Least Cost Routing is an example of a service provided to the whole user community, but is exclusively associated with networking. Therefore, it is not a system feature. Other examples of system features are Dial Plan, Direct Inward Dialing (DID), Directory, Emergency Transfer, and Direct Inward System Access. Dial Plan sets the high-level standard around which station extension numbers and dial access codes are assigned. For example, if the first digit is "9," the call is to a trunk group. Or, if the first digit is "4," the call is to a station. DID is the feature that allows callers outside the PBX to dial a standard seven-digit number and be connected directly to a station on the PBX without involving the attendant. Emergency Transfer is the feature that connects several predetermined stations directly to central office trunks if the PBX should experience a complete power failure. Direct Inward System Access is a feature that allows authorized callers from outside the PBX to have direct access to PBX services such as long distance calling.

Station features

Station features are features provided to individual station users on a station-by-station basis. Examples of station features are Call Hold, Call Forwarding, and Do Not Disturb. Call Forwarding and Call Waiting are familiar to most people because they have used them. Do Not Disturb provides a fairly unobtrusive signal (such as a very short audible "ping") to the station user for an incoming call and then immediately redirects the call to a coverage point such as a secretary or the Voice Mail System.

Many station features that are provided by the PBX could be provided by the station sets themselves. These features were first developed before the days of microprocessors when the only practical way to provide them was through the central call processor. With the advent of the cheap microprocessor and memory, many of these features are now provided by station sets. Call Hold and Speed Dialing are two examples. Important limitations on how many of these features can be implemented in the station set are:

- With analog technology, only one call can actually appear on a single line set at a time.

- The information bandwidth between the station set and the call processor is very narrow (it consists of switch hook flashes that last about a second each).

The list of options implementable in the set is short since the set itself cannot switch between or switch together multiple calls, and the signaling it can provide to the call processor is primitive.

This problem could disappear with Integrated Services Digital Network Basic Rate Interface (ISDN BRI). ISDN BRI provides two digital talking paths and a 16-Kbps signaling path to each station set. This allows local switching between two calls for local conferencing arrangements or call selection, and sophisticated signaling to the call processor regarding how to treat the call. Using this capability, an intelligent station set, such as a personal computer, can implement user-specific algorithms for call and information communications control. Applications for this technology include information sharing and exchange between multiple computers while the using callers are discussing the common information on a separate voice channel. Coupled with modern graphics packages, it facilitates a low-cost video teleconference where the video is limited to common viewing and editing of graphics.

In my opinion, for ISDN BRI to become a major player in station feature development, cost for this service will have to drop dramatically. There is a demand for these services and others, not invented yet, in the large and small business market, but the cost is too high to attract the mass market needed to stimulate such development. Economics are still in favor of implementing most of these features in the central call processor rather than the station set from a user point of view. It's not clear that central implementation is most economical from the point of view of cost of providing service. In other words, the ISDN vendors, for business reasons, are charging a premium price, well beyond the cost of providing service.

Messaging features

Messaging features are all those features designed to deliver messages between parties when one or both of them are not physically present. These began with the simple message waiting light that secretaries turn on to alert the boss that they have messages they have not collected. This has expanded to simple message features like Leave Word Calling, Message Desk, and Call Coverage Service. The current vogue feature in this group is Voice Mail.

Leave Word Calling allows a caller to leave a simple "call me" message with a telephone number at the push of a button. The message content is retrieved by the called party using a printer or display telephone. Sometimes the message contains the name of the caller which is retrieved by the call processor from the system directory using the called party's extension. The Message Desk feature routes unanswered calls to a common pool of message clerks who record the message much as a secretary would do (except not as knowledgeable about the called party's schedule). Call Coverage allows station users to specify where unanswered calls should be directed via System Administration. Call Coverage allows specifying a sequence of stations or station pools (e.g., Message Desk) to try in case the first or second choice is also busy.

Voice Mail provides much more flexibility than the other messaging services. First of all, it allows you to leave a much more extensive message with any telephone set because you can continue a monologue and say all you have to say, just as if the person was listening at the time. Secondly, it allows you to dictate a message at one time for delivery at another (say announcing an organization change just at the time it's effective). It also allows you to deliver the same message to a list of people at a specified time. Beyond this, it allows you to do most everything paper mail allows such as forwarding messages to others with annotations. Voice recognition techniques provide an avenue for authentication of the originator, much as a signature serves for paper mail.

Networking features

Networking features are all the features used to connect PBX users with the world beyond the free call radius of the PBX location. These features authenticate the calling privileges of the caller for placing the requested call, and select the route (trunks) to be used. They queue callers for facilities when all facilities are busy, and call them back when they become available. They control all access to Premises Based Private Networks, Local Exchange Carrier (LEC) Based Private Networks, Interexchange Carriers (IXC) Based Private Networks, LEC Public Networks, and IXC Public Networks (both circuit-switched and packet-switched). They implement and control network Uniform Dial Plans and Extension Number Steering (ENS). Using ENS, any particular extension number can be located on any PBX in a community of PBXs. Each PBX has to know which PBX the extension is on to route the call; therefore, administering ENS requires coordination between all PBXs within the ENS network.

Examples of features in Networking are Automatic Alternate Routing (AAR), Automatic Route Selection (ARS), Centralized Attendant Service (CAS), ISDN Primary Rate Interface (PRI) and Basic Rate Interface (BRI), and Main-Satellite. AAR and ARS provide automatic selection of the route that the call will take through the network. CAS provides for a group of PBXs to share a common group of attendants at one of the locations. Main-Satellite provides for a single PBX (Main) to do network routing for other PBXs (Satellites).

Control of network access and *caller authentication* are important functions of networking. Features supporting this are Facility Restriction Level (FRL), Traveling Class Mark, Code Restrictions, Authorization Codes, Barrier Codes for Direct Inward System Access, and Code Restriction. FRL provides levels of authorization whereby the services authorized are defined by the restriction level. Traveling Class Mark provides for transmitting this information on the originating party between PBXs along with the call so that appropriate restrictions will be applied at other locations. Code Restrictions allows defining what Area Codes and Office Codes are permitted to be called. Authorization Codes and Barrier Codes are numerical passwords used to authenticate the caller's identity. Together with other features, these are used to control telecommunications fraud as explained in Chap. 6, "Security and Fraud Control."

Data communications features

Data communications features are those features designed to provide data connectivity between data endpoints and ensure the reliability and integrity of those connections. PBX data features facilitate connection of endpoints using different communications protocols by providing protocol conversion. This includes provision of gateways to connect circuit-switched data to packet-switched data for connection to Local Area Networks (LANs) and Wide Area Networks (WANs). They also provide for multiplexing circuit-switched data endpoints onto T1 facilities as packet-switched data to reduce long-haul transport costs. They provide for connecting 3270-type terminals to cluster controllers through circuit-switched connections, thus eliminating the need for dedicated coax with the attendant inflexibility for moving the terminals. They provide permanent dedicated connections between data endpoints and host computers. This allows use of inplace house wiring connected to the PBX as though it were hard-wired between the two endpoints.

Data is often transmitted through station ports also used for voice communications. For example, establishing connections between FAX machines often requires voice communication between operators at each end to prepare the FAX machines to communicate. This can effectively control costs for receiving junk or inappropriate FAX messages. I do this with my FAX machine to avoid receiving junk FAX messages. Some station features, such as Call Waiting, interject tones or other voice circuit signals to alert station users of other calls. Such interruptions can damage or destroy the integrity of a data connection; therefore, data features provide for locking out these disruptive signals for data calls.

Data features also provide for 64-Kbps packet-switched data over ISDN PRI facilities. Using separate channel signaling for supervisory control of the channel, the entire 64 Kbps are available for user data. Similarly, ISDN BRI provides two 64-Kbps information channels plus a 16-Kbps signaling channel to a station location. The 64-Kbps channels are circuit-switched and may be used for voice or data. The 16-Kbps signaling channel is packet-switched and is used for supervisory signaling and may also be used for packet-switched data. This permits supporting two voice stations using the two 64-Kbps channels and several data endpoints using the 16-Kbps signaling channel from one ISND BRI port.

Efficient connectivity to host computers is also provided both by proprietary interfaces and using ISDN PRI. Both schemes are used to either circuit-switch data or packet-switch data to the host. These interfaces use T1 interfaces for the physical connection.

Some of the data communications features are listed in Table 4.1.

Call center features

Call centers are pools of agents serving a common incoming or outgoing call purpose. The most common form of a call center is an Automatic

TABLE 4.1   Data Communications Features

DS-1/T1 Interface	Multiplexes up to 24 data or voice channels onto each T1 interface.
EIA Asynchronous Interface Port	Provides for direct connectivity to devices using RS232 protocol.
Modem Pooling	Provides a group of shared modems for converting digital data transmitted internally to the PBX, using internal proprietary digital protocols to standard analog modem protocols for communication over analog facilities to distant modems.
LAN Interface	Provides a gateway for data transported internal to the PBX using proprietary digital communication protocols to RS232 protocol for connection to a LAN gateway.
PC Interface	Provides an interface between personal computers and the proprietary internal digital communications protocol for facilitating PC control of call setup.
Permanent Data Connection	Provides a dedicated connection through the switch matrix for data endpoints.

Call Distributor (ACD). ACDs are used to distribute incoming calls among agents with the required information or facilities to serve the expected purpose of the call. Examples of ACD applications are help desks, telephone order sales personnel, and insurance claims agents. Outgoing call centers are used to distribute the workload for placing and serving a large number of similar outgoing calls to a pool of agents. Examples of outgoing call center applications are soliciting orders for newspaper subscriptions, taking telephone polls to determine public opinion on some topic, and soliciting donations for charities.

ACDs are commonly used to serve callers calling "800" numbers. Typically, the 800 number is listed for customers to call for a specific purpose such as placing an order for merchandise, say photographic equipment. Agents receiving calls have information about current products for sale, such as price and availability, and are equipped to take orders and make commitments regarding delivery schedule. They are not equipped or knowledgeable enough to answer questions about problems with delivered products (customer service). Customers requiring that information are instructed to call a different number where agents are equipped to answer those kinds of questions. The point is that the business with the ACD knows what information and skills will be required to service the calls before they are received because the number the customer called was to be used for this limited purpose. If someone calls with a different question, they are directed to call a different number established for that purpose.

ACD agent groups are server groups, just like trunk groups. Therefore, the same statistical mathematics apply to them as to trunk groups, though the parameters are somewhat different. This means that to give high-quality service, a small agent group will be less busy, on the average, than a large agent group. Therefore there are significant economies to be realized in large agent groups.

What if the business volume doesn't justify a large agent group? Then there are a couple of options. You can have several 800 numbers, used for different purposes, terminate on the same agent group or overflow traffic from a small group to some other agent group, usually used to service different calls. The agent display indicates what number is being called, and hence what kind of service the caller wishes, before the call is answered. This requires agents and facilities that are more versatile than the single-purpose agent group. Many of the ACD features are designed to provide the required versatility for agent group sharing and to provide instant information to the agents pertinent to the purpose of the call they are serving.

Outgoing call centers often use computers to generate calls to a telephone list and distribute the calls to an agent when they are answered. The purpose of these calls are often very simple, such as "Are you currently subscribing to 'The Rural Rag' and, if not, would you like to take advantage of our special offer?" Information about the called party, such as name and address, is often provided to the agent when the call is distributed to them so that they can appear more personally interested in the potential customer.

Outgoing call center groups can usually handle more calls of equivalent length (higher agent occupancy) than ACDs because the call center has complete control over when the calls are placed. This eliminates the random process of uncorrelated incoming calls since the call generator only places calls when agents are available to service them. To place calls at appropriate times, the call generator must have current information regarding the availability of agents.

Management of call centers includes the management of people doing production-type work. Call center features provide data on agent performance for each agent. Some of these measures, such as average work time per call, number of calls served, and total time each position was staffed, are standard. Others can be customized to the special needs of the particular application. Examples are number of sales made, total volume of sales, and number of specialty items sold. How these are weighted for agent evaluation is a business specific decision. However, the PBX Call Center Features provide significant data to facilitate these management functions.

Examples of Call Center Features are shown in Table 4.2.

TABLE 4.2   Call Center Features

Call Distribution	Distributes calls among agents according to user-selected algorithms such as Most Idle Agent or Circular Hunting.
Queueing	Places calls in a queue (or line) until an agent is available to service the call. The queue may be First In First Out (FIFO), Last In First Out (LIFO), or by call priority—where priority is determined by what number the calling party dialed. With ISDN and ANI, priority determination can be augmented by where the call originates from.
Agent Force management	Provides data on agent performance such as number of calls served, average after-call work time, agent log in/out time, and average call holding time.
Service Observing	Allows agent supervisors to monitor on-call performance of selected agents to verify proper call handling procedures without the agents knowing when they are being monitored.
Split/Group Overflow	Provides for overflowing calls from one agent group/split to another when the first group is busy. Sometimes this is done from the incoming call queue when the time in queue for a call has exceeded a predetermined threshold.
Call Vectoring	An enhancement to Split/Group overflow that allows the PBX user to write a program to define the call distribution and group overflow logic for redirecting calls for service. These programs can redirect calls to different agent groups on different PBXs based on parameters such as time of day and the remote agent group occupancy.

Specialized business features

Specialized business features are features developed for a specific business sector rather than for the business community in general. Two specific areas where this has occurred are Hospitality Service for hotels and motels, and Multiple Tenant Services for companies that lease buildings to multiple tenants. Hospitality Service provides for tracking the status of rooms, assuring that calls are only made from rooms with appropriate billing authorization, and special guest services. Multiple Tenant Services provide for several tenants, served by a common PBX, to have it appear functionally as though they each have a separate PBX.

Hotels and motels with many rooms must accurately know the status of all their rooms. Room status is information regarding whether the room is rented, and if not, whether it is ready to rent or needs some service. Typical status states for rooms are:

- Ready to rent
- Guest checked in
- Needs cleaning
- Room cleaned
- Guest checked out
- Needs maintenance

Accurate information on room status allows renting to higher occupancy levels because the front desk confidently knows that rooms are available and the risk of irritating guests by renting an uncleaned, defective, or occupied room are reduced. Hospitality Service features provide for changing room status by housekeeping and maintenance staff by dialing a status code when they perform the needed services. These status change calls can also be used to track service personnel progress.

Room status is also used to control calling privileges from guestroom telephones. If a room is not occupied, outside calls and toll calls are restricted. This is just another way of controlling toll fraud as discussed in Chap. 6, "Security and Fraud Control."

Special guest services include features that track guest requests such as Do Not Disturb and Wakeup Calls, and provide for consolidating guest bills from throughout the facility via telecommunications. All of these services are unique to the guest lodging industry and have been developed to serve that specific industry segment.

Some vendors have chosen to provide these services using equipment and software that they have developed. Some are directly integrated into the PBX call processing software and some are provided by adjunct processors. Adjunct processors were used by AT&T for many of these services with its Hospitality Services adjunct to the Dimension® PBX and Dimension System 85® PBX product lines. The Hospitality Services adjunct line was discontinued. Other venders, such as Northern Telecom with its Meridian® PBX line and IBM ROLM, provide an interface between their call processor and other vendors' products for integrating call processing functions with the hospitality services provided by vendors specializing in that business. The experiences of these vendors in providing these services seems to support the principle that businesses specializing in particular products and services do a better job at providing them.

Multiple Tenant Services allows a PBX serving a location to be configured to appear as several separate PBXs (virtual PBXs) to the users. This means that each virtual PBX has its own trunk groups, attendant partition, and extension numbers. This facilitates each tenant managing their own telecommunications costs independently by optimizing network services and access authorization for their business.

Extension numbers in one virtual PBX cannot call extensions in another virtual PBX by dialing just the extension number, but must access a central office trunk and call the Direct Distance Dialing (DDD) number. A single attendant position is allowed to serve more than one partition (virtual PBX), or each partition can have its own attendants. This feature allows building owners of multiple tenant facilities such as airports, industrial parks, and medical centers to provide separate facilities and services to tenants with different requirements without purchasing separate PBXs, thereby providing a cost savings both in purchase and operation of the telecommunications facilities.

Both Hospitality Service and Tenant Services were developed to serve specific important market segments. As such, they have limited application and aren't as well known as many of the other features discussed above.

Applications Processors

Applications (Adjunct) Processors are systems that surround the PBX and provide facilities and services in support of the basic PBX service. These processors provide additional features and services that users require to enhance the Basic PBX service. Many features and services provided on Applications Processors are also offered using the PBX call processing processor; however, Applications Processors implementations are often significantly enhanced. For example, System Management is often offered via Applications Processors. It is also offered using just the PBX call processing processor. The difference is that the version offered using the call processing processor is very bare-bones, sometimes with an almost abusive human interface. Conversely, as offered on Applications Processors, it is very user friendly with screen-oriented help, standard templates for station set administration and trunk administration, sophisticated consistency checking, and automatic administration change event logging.

The choice between implementation in adjunct processors or on the basic call processor is driven by several considerations. Features and services developed on adjunct processors are developed independently from the PBX, allowing the vendor to speed the entry of both into the marketplace. Independent separated products connected through a simple well-defined interface are easier and faster to develop and evolve than highly integrated products that share common hardware and contend for common computing resources. They can be kept relatively immune to recurring changes in the PBX's software and hardware. This simplifies the design, particularly the software, and ongoing product evolution. Choosing an appropriate operating system for the adjunct processor (such as UNIX® or DOS®) can make the applications software development relatively independent of the hardware base. Thus, a simple migration path exists as processor products improve in size, price, and performance.

Finally, adjunct processors serve the valuable function of deloading the call processor from the work of handling these applications. This is especially important because the processing resource requirements for call processing and many enhanced applications are very different.

Call processing software is required to respond to many call processing stimuli (requests for processor service) within a few milliseconds. The processing required usually is very simple and is completed in less than a millisecond (actual numbers vary with the total call processing capacity of the PBX). Call processing is characterized by many requests for service that must be served very quickly, each needing just a small amount of processing.

Conversely, many adjunct processor applications are characterized by relatively few requests for service that each require several milliseconds of processing. The difference in response time requirements results from call processing interacting with other fast, computer-controlled switching machines while adjunct processor applications are interacting with impatient, but slower, people or are doing non-real-time batch processing such as CDR record processing to generate customer bills. The difference in the processing job size results from call processing stimuli requiring simple updating of a state table or adding a tone to the connection while adjunct processor applications are often performing database operations.

What this means is that the processor cannot perform both functions efficiently because optimizing for each is so different. It's similar to trying to choose a single vehicle for family vacations and hauling gravel. It can be made to do both, but at least one of them is not going to be done as well as if it didn't have to do the other. Adjunct processors allow applications with similar processing requirements to be hosted on a common machine while being separate from other applications with dissimilar processing requirements.

In early implementations, the APs were of the minicomputer family. These applications were very complex and required considerable memory and processing power; therefore, these applications were often each hosted on a separate AP. End users wanted processors in which multiple applications could be coresident because the applications processors were often idle. *Coresidency* means that several applications reside in a common processor, and the user can select each application in a simple manner. Also, in a coresident environment, the applications may need to share common information in the system memory.

Another strong end-user requirement was for a multiuser, multitasking environment. Multitasking means that more than one function can be in operation simultaneously. For example, an application which processes call detail information can be run at the same time that an application which provides a system directory is running. In a multiuser system, several terminals can be operated at the same time. Coupled with the multitasking capability, the functionality permits one user to operate the call detail application while another is running the directory application. The multiuser capability also facilitates supporting multiple simultaneous system administrators on larger systems where the level of administration activity is more than one person can handle.

Previously, only minicomputers had the necessary processing power, memory capacity, and disk storage capacity required to satisfy these requirements. Now, with the advent of low-cost powerful multiuser, multitasking microcomputers, the market for adjunct processors is shifting in that direction.

There are a variety of applications available on adjunct processors targeted to the telecommunications end user. These applications are grouped into groups of similar functionality to facilitate understanding.

The first group of applications to be considered is Messaging Services. The purpose of this group is to permit the telecommunications end users to better manage the flow of information to and from them.

Message Center.   Allows a centralized message center to answer and record messages for the client's unanswered or redirected calls. The recorded messages can be directed to display telephones or to conveniently located printers.

Voice Mail. Provides facilities for recording, editing, storing, manipulating, and sending voice messages. These functions can be used by callers to telecommunications system end users or by users who are subscribers on the voice mail system. Voice mail service is much more sophisticated than that provided by answering machines.

Electronic Mail.   Facilities for sending, receiving, editing, storing, and redirecting electronic mail. Electronic mail is that which originates on or is directed to intelligent (usually computer) terminals. It is computer-generated text, graphics, and spreadsheets. Electronic mail provides a very efficient way to transmit complex documents electronically. Its principal advantages over FAX are that it produces documents at the receiving end whose printed quality is largely determined by the printing facilities; documents that can be edited without reentering them; and faster, lower-bandwidth transmission.

Leave Word Calling.   The facility to permit the station user to push a button and direct a message to a display telephone or conveniently located printer to have the called party return the call.

Unified Messaging.   Provides a message-waiting alerting capability that tells the user that messages are waiting in the system from one of the several messaging systems (voice mail, electronic mail, Leave Word Calling, etc.).

The second group of application processor functions to be considered is System Management Services. This is the functionality group that allows end users to better manage their communications systems. These functions are discussed more fully in Chap. 5, "System Administration, Maintenance, and Reliability."

Terminal Change Management.   Provides capabilities for making administrative changes to the system for controlling terminals. This includes adding terminals, changing terminal type and functionality, changing access permissions, moving terminals, and deleting terminals.

Facility Management.   Provides capabilities for making administrative changes to trunk-type facilities.

Automatic Trunk Test Measurement System.   Provides facilities to test trunks for adequate transmission and noise levels on a scheduled basis.

Network Management Systems.   Provides the capability to receive alarms, diagnose systems, provide network displays, and assist the network manager in ascertaining how well the PBXs and network systems are performing.

The final group of applications processor features are those called Information Systems Management. They allow the end user to better manage their telecommunications service.

Directory.   Provides an electronic directory (that can be printed into a corporate directory), usually listing name, telephone number, location, and the individual's access to messaging systems. The directory functionality is also used in conjunction with Message Center Service.

Cost Management.   This includes the Station Message Detail Recording and Reporting functions which are used to record selected data on calls so that telecommunications costs (such as toll bills) can be "fairly" allocated to the user community. What is "fair" is determined by the algorithms used to convert CDR records to telecommunications bills. These applications receive data on each call processed by the PBX and process the data to rate the call—that is, to calculate the cost of the call. The data from the PBX also includes information identifying the billable extension or account code associated with the call. These data are combined to produce detailed bills to the responsible organizations.

ACD/MIS.   Provides a management information system that supports the Automatic Call Distribution functionality of the PBX. The MIS system delivers standard reports (and permits the user to customize reports) which provide the administrator information about the performance of the ACD system and the system agents, and facilitates system performance improvement.

Office Communications Systems.   Provides packages of office-oriented software to facilitate information movement and management. For example, one popular package provides PC management of incoming and outgoing call volumes.

Originally, Applications Processors connected to the PBX over proprietary data links. Since the vendor was expanding the price/performance of the overall system through the use of an external processor, proprietary linkage was a natural strategy. However, from the end-user perspective, it was a negative influence. Suppose, for example, that Vendor A sells a PBX with all the desired functionality at the best price. However, Vendor A's voice mail system is not state-of-the-art. Vendor X's voice mail system is top-rated and competitively priced.

What is the user's solution? An attractive alternative is to try to reverse-engineer the link from Vendor A's PBX and develop a blackbox interface. That alternative is costly and risky.

The success of the call management service market has encouraged many vendors to develop applications running on a variety of hosts. Vendors have been unwilling and unable to duplicate these applications. Consequently, PBX vendors have been forced to open the interface to the PBX to permit easy, two-way communication with the wide variety of hosts. The solution developed is the Switch-to-Computer Link. Some variations are called Response/Status Links (RSLs).

A typical arrangement of PBX, host computer, and SCL is illustrated in Fig. 4.1.

The functionality illustrated is typical of a company that operates a service to provide information to callers through an ACD. The information service might be a help desk, customer order status, or billing information. In this case, "dealer location" service is provided by the agent. That is, the agent provides the caller with information on where the nearest dealer of a product is located.

The incoming caller is connected to the PBX, which gathers calling party information from ISDN PRI or ANI messages associated with the call. In processing the call and directing it to an agent (or queuing the call as appropriate), the PBX strips off the calling party information and sends it to the host over the SCL. The host processes the calling party information and searches the database for the location of the nearest dealer to the calling party location. Through the application, it

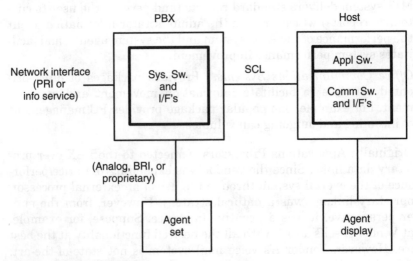

Figure 4.1 Typical PBX/host interface.

delivers the nearest dealer information to the agent display. The voice and data path to the agent can be an ISDN BRI link, a proprietary voice/data path, or separate analog and digital paths.

In practice, the SCL encompasses the interfaces, protocols, and the digital link. It can be provided in two ways. In the direct-link case, the PBX contains the hardware and the application program to manage the information transfer protocol. However, some vendors choose to isolate that function from the PBX and provide a standalone gateway. The gateway consists of a separate application running on a stand-alone processor. The feeling in the industry is that the gateway approach will be used more to isolate the SCL from changes in the PBX and hosts and, more importantly, to provide the capability for a single PBX to access multiple hosts.

As illustrated in Fig. 4.1, there is a need for transport of information residing in the network to the application. Today, local exchange and interexchange carriers make available the calling party number through Automatic Number Identification (ANI) and Dialed Number Information Service (DNIS). ANI provision by the carriers has been challenged in court as a violation of privacy. A court in Pennsylvania recently ruled such provision violated the state's telecommunications privacy law which is modeled on a federal law.

DNIS is useful in a multicustomer/multifunctional inbound call management system. In such systems, the caller dials a specific 800 number group to obtain some type of information service. Other callers call a different 800 number to obtain different information. Yet a third caller dials another 800 number to obtain yet a different type of information. The network, however, routes these calls to the center over a single group of trunks. DNIS permits the system to identify what type of information the caller is seeking, direct the call to the agent group most appropriate, and provide information screens to the agent pertaining to the type of information sought by the caller. Or, the application may direct the call to a group of agent providing multifunctional service and provide the agents with the appropriate screens to serve the call.

Referring to Fig. 4.1, the system has a number of discreet characteristics:

- The switch application program usually handles the routing of the data message and always manages the voice switching function.

- Industry standard open protocols, such as Q.921 and Q.931, are used to ensure transport of the required information across the switch/ host interface.

- The host application processes the information from the switch, calls up its database, and directs an output to the application user.

- The links from the switch to the host, and from the host to the application user, generally are relatively low-speed. The switch/application user link generally transports voice and data. As previously mentioned, the ISDN Basic Rate Interface is an ideal transport system. Proprietary interfaces, such as AT&T's DCP, can also be used.

SCL systems find a market in the telemarketing (inbound and outbound call management) sector. In any application where rapid call redirection is required, or where access to specialized information is required to serve the caller, SCLs have a use. They can be effective in routing calls serviced by voice answer systems with automated attendant functionality. The operations of help desks, customer service organizations, and financial services businesses are typical of those that can benefit.

While it is early in the market life of these products to accurately project savings, some industry data is available. As background, an SCL system can start at a low-end $40,000–$50,000 price and move to the hundreds of thousands of dollars range if a new PBX, major software development, or a large agent population is required.

One user group has released some averages based on the combined experience of their several companies. They found that improvements in agent productivity was in the 17–20 percent range. Depending on the specific application, this can translate to a 5–15 percent reduction in call handling time. In a large telemarketing center handling 40,000 calls per day, this reduction translates to a 50–150 person-hour per-day savings. Assuming a $10/hour cost, the savings is $500–$1500 per day. On a $200,000 investment, payback is reached in only 4–12 months.

Conclusion

Applications are all the features and services that the modern PBX provides on top of basic calling capability. Features have been developed for many user communities, beginning with the basic office telephone user. Features for the basic business user are designed to make it easier for them to place calls, leave messages, receive (multiple) calls, provide telephone coverage (e.g., messaging services) for when they are unable to answer their calls, and reduce the cost of toll calls. Other features have been developed for system administrators, maintenance personnel, data communications, and people with special telecommunications applications. Most features are intended to make the PBX a better resource for meeting corporate objectives, including becoming a principal revenue source and reducing the cost of providing telecommunications service.

Most businesses can benefit from many of these features, but virtually no business can benefit from all of the features. When choosing a

PBX, it is important to determine how any particular PBX's features fit the needs of the particular enterprise it is to serve. The needs considered should go beyond just the functions the business now performs. It should include creative ideas on how the telecommunications service can change the way business is conducted to improve revenues and reduce costs. If the PBX purchase doesn't serve these purposes, then the management needs to look seriously at how payback will be achieved from the expenditure. Certainly, this is not the only criterion, but it is a very important one.

Make sure the PBX you purchase really serves your needs. It is easier and wiser to purchase a PBX with the features you need than to purchase one in hopes that the vendor can be persuaded to add the needed capability!

5

System Administration, Maintenance, and Reliability

The Private Branch Exchange, whatever its technology and architecture, creates the requirements for administration and maintenance services. These services ensure that the system makes its connections flawlessly and acts reliably. There is no industry standard definition for these terms. Collectively, they refer to operations required to run the PBX as a facility for the station and terminal users that are connected, and to control and allocate the incurred costs. Administration, as used in this chapter, refers to the management of parameters used to assign stations, terminals, and facility groups; to choose options from among those offered for the system as a whole, for individual station or terminal users, and for individual facilities and facility groups; and to allocate the incurred costs.

Management of parameters includes:

- the establishment of policies for the allocation of system resources
- translating these policies into specific parameter selection
- entering the selected parameters into the system
- establishing routine performance and usage monitoring procedures to verify proper implementation of the policies
- change procedures to respond to move requests and correct inconsistencies between system performance or use and the established policies
- publishing of those policies and parameters in appropriate form such as telephone directories

Policies for the use of system resources address the questions of who is permitted to do or use what resources under what conditions and for what purposes. This includes items such as what type of telephone equipment is appropriate for various job classifications. Who gets access to long distance service and at what times of day (a company may decide to allow long distance access only during normal working hours) and via what facilities (trunks)? Who, if anyone, may use Direct Inward System Access (DISA), a feature that facilitates access to PBX services via an outside line if a proper access code is dialed? (Failure to manage this feature properly can cost a company its business in unauthorized long distance bills.) What area codes and office codes are permitted to be called? (How much business do you expect to do in Columbia?) Who is allowed to change the control parameters under what conditions? These and many other parameters must be managed to assure that a PBX is a company asset rather than a liability.

Cost management and allocation refers to establishing policies for incurring telecommunications service bills and equipment costs, measuring the costs, and allocating them to the appropriate billing entities (individuals or organizations). Measurement of costs incurred by billing entities is usually based on equipment assigned and calling traffic (both local and long distance) as measured by the Message Detail Recording System to be discussed later.

Maintenance includes all functions required to detect, track, isolate, and repair problems detected in the PBX system or attached hardware. Some of them are:

- monitoring error logs for patterns of failures
- responding to user-reported problems
- monitoring and responding to alarm conditions
- performing and evaluating periodic and demand testing for failure conditions, and isolating identified problems
- problem analysis
- dispatching repair forces
- maintaining accurate inventory records of spare equipment, replacement parts, and cable pairs

One principal measure of system performance is *system availability*. This means when the system is called upon to perform a service, such as complete a call or record a message, what is the probability that it will be ready to fulfill the request? (This will be discussed in detail later in this chapter.) The fundamental purpose of maintenance, whether preventive or corrective, is to maintain satisfactory system availability at reasonable cost. Failure to meet user expectations for system availability will create more unwanted attention quicker than any other service deficiency!

The following sections give a detailed view of system administration and maintenance.

Administration

As described above, administration is a broad term that encompasses all the housekeeping tasks associated with the proper management and programming of the system. The administration system deals with the establishment of telecommunications policies—system parameters—to carry out those policies, and monitoring the system performance to assure that the resulting performance is satisfactory and consistent with the established policies.

Establishing sound policies begins with an understanding of the business enterprise and the role the telecommunications system is to fulfill within the enterprise. Is this a multilocation company requiring a private Electronic Tandem Network or is it a single location company? Which lines or groups of agents generate revenue (say, a sales call center) and which are an expense (a help desk)? Which employees or types of employees require access to long distance services? Which job functions could be made more efficient with enhanced functionality telephone sets? What are the typical calling patterns for toll calls and where can toll bills be reduced using Wide Area Telephone Service (WATS), tie trunks, or foreign exchange trunks? How many WATS trunks, tie trunks, or foreign exchange trunks is it economical to purchase? What areas of the country or world do users have no business calling at company expense? How will calls be treated when the principal called party does not answer? When the system fails, how fast should service be restored (faster repair time costs more in labor, parts, and initial investment)? Answers to these and similar questions form a framework for choosing among the many facilities and services available through the PBX.

Administration parameters

Administration parameter categories. Administration parameters can be divided into three categories, depending on their scope of control. *Scope of control* refers to how much of the system is affected by changing that parameter. The greater the scope of control a single parameter has, the greater the potential effect of a change in the parameter. For example, terminal button assignments affect a single user while Code Restrictions affect the whole system and the long distance bill. The parameter categories are Systemwide, Facilities, and Individual Station User and Terminal parameters. These are described in the following paragraphs.

Systemwide parameters include items such as dialing plan (two, three, four, or five digit; leading digit assignment; network access digits; etc.), cabinet configuration, and Facility Restriction Level (FRL) access privileges. These are parameters that define the system configuration and determine what features and services are provided and how they will be accessed.

Facilities are items that are accessed as a group by stations or other facilities such as trunks and host computer ports. Facility and facility group parameters are items such as trunk type, supervisory signaling method, trunk group to which a trunk is assigned, number of trunks in the group, whether calls are blocked or queued when all trunks in the group are busy, etc. These are parameters that define what shared-use facilities are available, how many are provided, and how they are physically and logically connected to telecommunications system users and external systems.

Individual station or terminal user parameters include items such as the extension number, the physical port to which the station is connected, features that can be accessed, facility restrictions, and telephone button assignments. These are parameters that define what station or terminal equipment is used, how it is connected to the system, what feature and service access privileges are allowed for the terminal or station user, and how station or terminal hardware is logically linked to features and services (e.g., station set button assignments).

Parameter selection.   The administration parameters are specified by the end user. Often a "station review" is conducted by the vendor to decide what the parameters should be. During the station review, the system administrator, the system operators or attendants, and individual users are asked preferences in service and equipment.

The system administrator provides general system information and policy guidelines. That includes the preferred numbering plans for terminals, network access, and special service systems. The administrator also specifies the number of trunks, trunk groups, and permissions granted for access. They provide traffic information on busy hours, weeks, and months, and the growth information that is so important to the proper engineering of the system. Engineering for growth is essential to avoid purchasing a "white elephant." Many PBXs have a maximum capacity beyond which they cannot grow. If this breakpoint falls below the expected final growth size, a completely new system will be required when that limit is reached. Conversely, if a much larger system is purchased than is needed, it will be a much larger initial investment than is necessary.

Attendants frequently have a wide latitude in the specification of functions to be provided. Most attendants are quite experienced in managing their telecommunications system. They often provide outsiders with their first impression of the corporation since they answer most calls from outsiders to The Company (Listed Directory Number calls). Their job, well done, can be a positive influence on the business. They are often instrumental in selecting the type of messaging adjuncts provided with the PBX since they have a practical understanding of how those adjuncts would fit into current practices and personal preferences of key people for taking and distributing messages. Usually, they review the engineering information that establishes the number of attendant consoles provided. They also are very useful in establishing the type of directory (electronic that is printed to paper copy) that will be used.

The station users usually have the latitude to specify button assignments for their terminals, the call coverage paths to be used, their access to adjunct features, and other comparable choices. Frequently, a common terminal with common feature button assignments will be selected for groups of people within the location. This would apply, for example to a group of engineers at an R&D company. In these cases, terminal administration is replicating the common assignments, adding the extension numbers and user names, and identifying the physical port where each terminal is attached. Most often, the type of terminal that users are provided is a corporate choice rather than an individual choice.

Parameter initialization. The administration data to be used at cutover is loaded into the system in bulk during initial installation. Often there is a period ranging from days to weeks between when the cutover administration data is supplied and when cutover occurs. This results in a backlog of changes to the bulk-loaded administration parameters that must be entered before the cutover is complete. These changes are the accumulated moves and changes that routinely arise in the normal conduct of business. They result from such things as employees entering and leaving the business or changing their work location. Subsequently, changes are made periodically (daily, or perhaps weekly or monthly) depending on the activity associated with the system. The rate of administrative changes can be as high as 6 percent per month in active systems. Thus, the initial administration of the system could be entirely redone in less than one and a half years.

In Chap. 3, the architecture and design of the modern PBX is presented. In the architectural model, part of the system memory is reserved for the storage of the system administration information. That reserved part of the system memory is called the translation memory because it stores the information (database) to translate the choices that have been made into specific parameters to control actions for any situation such as how to process a call. This is the database that is controlled by the administration system.

Operational administration systems

The administration system provides the actual tools for changing administration parameters, and provides the interface between the administrator and the administration database. This interface is active while the database is being used for processing calls. Consequently, it is extremely important that the integrity and consistency of the database be maintained while it is being changed to maintain the integrity of call processing. To achieve the required database integrity, the administration system must perform validity and consistency checks on data entered and prevent erroneous data from being accepted. No data validation system can catch all errors (or else the system already knows what the data is and doesn't need it to be entered). The characteristics of the residual errors that the administration system cannot prevent will determine what the human responsibility must be for data validation. The administration system must assure that a meaningful set of parameter changes is entered before implementing any of them. For example, it cannot allow a trunk to be used before it is assigned to a trunk group.

The administration system also must provide an interface that a reasonable human can understand. It might not be user-friendly, but it shouldn't be user-hostile! An interface becomes user-hostile when users can't easily interpret current administration data or format new data. For example, it becomes hostile if users have to reduce administration changes to setting bits in selected memory words.

Hostile interfaces lead to administration errors. Data in the database is not easily understood with these types of interfaces; therefore, obvious errors are missed by the administrator. User-hostile administration interfaces encourage administrators to make subtle errors that affect system performance and are difficult to find. They also feed a cottage industry of administration specialists whose principal skills are in divining the causes of strange system behavior resulting from administration errors or making the administration changes more reliably than the casual user. It also can affect the usefulness of a PBX to a business. More than one business does not take advantage of many PBX business features because the administration change process is so difficult and unreliable.

The administration system must be capable of making parameter changes while the system is operating. The PBX is expected to provide service twenty-four hours each day, so it can't be shut down for administration parameter changes. This requires database integrity while the changes are being made (as described above), and real-time allocation of resources between call processing and system administration. For call processing and system administration to access a common database, they have to access common memory; therefore, there is contention for common resources. This can be memory contention through use of Direct Memory Access (DMA), or processor contention as they use the common call processing processor to access the common memory. In both cases, there is real-time contention for processing resources and a possibility of slowing call processing. In most modern PBXs, contention effects on call processing are minimized (but not eliminated) via a priority scheme giving call processing higher priority than system administration.

The administration system must provide a way for people to read and interpret the administration parameters that the system is using. Reading and interpreting administration parameters is part of troubleshooting and is often the first step in making changes (before you change something, find out what the current value is and verify that is what you expected). Also, after having made a change, it is common to verify that the change has been entered correctly by reading it back out.

Beyond these basic capabilities, administration systems available today provide many tools to reduce the work required to perform these functions accurately. They also provide ongoing monitoring of system performance to identify and correct problems before they have a significant effect on the user community.

Administration system architecture.   The form of the administration system varies from vendor to vendor and from system to system. First, the task of managing the administration database may be handled by the system processor. This is characteristic of smaller systems. This type of architecture requires that wholesale translation changes be made during periods of light traffic. The system processor will interleave the processing of translation data with its call processing responsibilities. If calling volumes are heavy, the processor may not accept translation changes at all or may handle them at low priority, resulting in slow response to administrative commands.

Alternatively, the system architecture may be such that the bulk of the work for translation changes is handled by a separate processor in the processor complex. The separate processor can assume responsibility for managing the human interface with the administrator, providing data entry screens, and performing validity checking against existing data. It can "bundle" a set of related changes to assure that they are made simultaneously. It can schedule changes to occur at times consistent with, say, organizational changes or moves. Because of the costs associated with a separate processor and the software to run it, this arrangement is usually used for large systems where the rate of change is expected to be high. Recent developments are resulting in the cost of these systems going down and thus becoming economically practical for smaller systems. Though most of the work is handled by an adjunct processor, the call processing processor usually becomes involved in actually changing the values in its translation memory.

Many systems use a split scheme in which much of the processing is delegated to a processor external to the PBX. In such systems, the interaction with the system processor is minimized. Often the split of functionality is chosen to build upon an existing system included with the internal processor where much of the data validity checking is already implemented. This means that changes to the administrative database can be done at any time except during unusually heavy traffic periods.

The terminal configuration used as the person/machine interface for administration depends on the architecture of the system. This is usually a tradeoff between cost and functionality, especially in the friendliness of the human interface. Where the PBX system processor or an auxiliary processor in the PBX processor complex handles administration, a dumb terminal will suffice. Sometimes, a unique terminal usable only with the specific PBX system is employed. Where processing is distributed outside the PBX with minimal system processor interaction, a smart terminal, micro-, or minicomputer is used. The bulk of the processing is done external to the PBX and the translation changes are downloaded to the PBX.

The interface between the PBX and the administration system (or terminal) generally is closed or proprietary. Vendors take that approach to preserve their administration systems as value-added devices. Most vendors offer a low functionality terminal through which the customer can process administration changes, and a high functionality system that offers attractive price/performance.

The human interface.   Whatever the architecture and terminal used, the user must consider the translation format required. When dumb or unique terminals are used, the person entering the translation data may be required to format the data in a form directly usable by the PBX (set bits or fields in memory words). This can result in time-consuming, tedious work. The change of a multibutton, high-functionality electronic set may require 10–20 individual translation changes. This translates into an ongoing operational cost for the PBX that can easily dwarf the initial cost of providing a robust and user-friendly administration system.

To ease translation work, the processor (either internal to the PBX or in an external adjunct) should provide the user with screens requiring English language inputs and provide help and tutor windows. These screens should be task-oriented where the tasks cover the scope of jobs (work orders) to be done. For example, a screen should be made to move an extension from one location to another. Another screen would add or delete an extension. Yet another one would add a trunk to a trunk group. The system should provide validity checking of data as it is being entered into the screen, and not wait until the system translation database is being changed, to discover the error. Delayed error checking requires multiple human interventions to correct errors, making errors more costly to correct. They also can substantially increase the time required to make scheduled bulk translation changes, upsetting planned schedule coordination with other facilities involved with the changes. The processor should be programmed to translate the input to the formats that are usable by the PBX, rather than using an error-prone manual system.

An interesting approach taken by several major PBX vendors is to provide an administration system adjunct that contains a "mirror image" of individual PBX databases. Often these mirror-image databases are created by uploading the translation database from the PBX. It is important not just to maintain separate databases and then have the same changes made to both them. Something will always happen to get them out of synchronization. This could be a system hardware failure, while the changes are being made, or human error.

Administration systems may be designed to be at a centralized location. All additions, deletions, and changes to the administrative database are done in the adjunct. The interface to the user is screen-oriented. Some of these systems can be run in a multiuser mode; that is, changes can be input by more than one person at a time. They are used on very large systems where the rate of translation changes is more than one administrator can handle. At an appropriate time, the adjunct calls up the individual PBX and downloads the database changes. This approach is very useful in a multi-PBX network arrangement where a change in one PBX might trigger the need for changes in other PBXs. The larger and more sophisticated of these are very expensive and can only be justified for large, complex systems.

Administration system functions

The functions of administrative systems can be broken down into five categories; they are management of the system, terminals, facilities, traffic, and cost. These are illustrated in Fig. 5.1. Of these, system, ter-

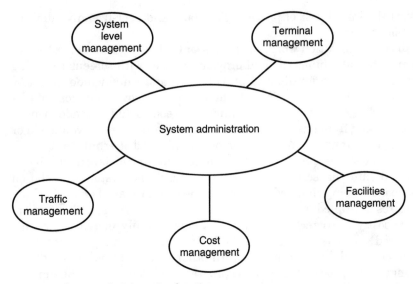

Figure 5.1   System administration functions.

minal, and facilities management are baseline services. Without them, no administration of the PBX is possible. Without Traffic Management, economic and cost-effective management of facilities is not possible. In other words, you can make the system run without Traffic Management, but the cost for leased facilities and the grade (quality) of service they provide are only as good as your guess for traffic levels. Without Cost Management, the real costs of using the telecommunications system are not known or managed by those who can effect necessary controls to prevent resources from being wasted.

System-level administration

System Management (system administration) cares for the housekeeping functions that directly affect the operation of the PBX system itself, regardless of the terminals, trunks and special services. Table 5.1 lists typical System Management parameters.

Many of these parameters are selected at installation time and are never changed, unless there is a major reconfiguration. Others, such as whether DISA is allowed or whether to use Automated Attendant Service, may be changed because of operational experience. Though these parameters are fairly stable, they must be managed in the sense that the effects of the selected values on the system and system costs must be monitored and reviewed to assure that they support the business

TABLE 5.1   System Administration Parameters

Security codes
System hardware information (cabinets, carriers, etc.)
Single or duplicate common control
Availability of cache memory
DTMF senders and receivers
Tone plants
Intercom records
Time-of-day (system clock)
Dialing plan
Classes of service
Directory database
Attendant consoles

goals. For example, Automatic Attendant Service may reduce operational costs (you don't have to pay attendants), but if business customers take their business elsewhere because they can't talk to a real
person, then lost revenues may far exceed the savings in staff. Similarly, DISA may be a useful and convenient feature for people who
work outside the business facility to access the low-cost corporate long
distance facilities, but if the access procedures and codes become
known to the wrong people, a company may find itself paying for
telecommunications for drug deals or for foreign nationals to phone
home.

Terminal Management

Terminal Management (terminal administration) provides the software
that is associated with system terminals (e.g., telephone sets). Terminals can range from simple zero-button analog telephone sets, through
complex multibutton digital sets with displays, to non-telephone-type
terminals, including computer equipment.

Terminal parameters completely define each individual terminal to
the system. This includes the terminal type, extension number, location (where it is attached to the PBX), assignment of all buttons on the
set, user privileges, other extension numbers that may appear on the
set, the user's name, how calls that are not answered or find the terminal busy should be treated, etc. Table 5.2 lists some terminal administration parameters. This is just illustrative, since the list of
parameters is extremely long.

This is the most dynamic part of System Administration. It is the
functionality used for any telephone installation, move, change, or
deletion. Most new products for simplifying System Administration
are directed at this area because about 95 percent of System Administration change activity is here.

TABLE 5.2   Terminal Administration Parameters

Virtual location of terminal
Physical location of terminal
Extension appearance
Name of user
Terminal type
Button assignments
Call coverage paths
Hunting
Personal central office line
Ringing preferences
Class of service

Facilities Management

Facility Management (facility administration) is similar to Terminal Management, except that it deals with trunk facilities and access to networks, hosts, etc. These parameters completely define both the facility groups and the ports within the facility group. For example, a trunk group is defined by the trunk type (tie, FX, WATS, etc.), the supervisory signaling type, the group size, the treatment of call attempts that overflow the group, dial access code, queueing treatment used, trunk group access restrictions, etc. Trunks are defined by what trunk group they are a member of and their physical port location. Table 5.3 lists some of the parameters. Again, because of the lengthy list of available parameters, those shown in the table are illustrative only.

Facility Management features are separated from the Terminal Management features because of the potentially large cost impact of these parameter choices. Many customers will allow access to terminal administration capabilities to a variety of corporate personnel, but will restrict access to facilities administration because of the potential cost impact. The level of facility change activity is much lower than terminal change activity; therefore, fewer people are required to keep up with the change activity.

Carrier-provided access facilities (such as trunks) are billed in a variety of ways. Some trunks are billed almost entirely on how much they are used (Central Office trunks). Others are billed on a combination of how much they are used on a group basis, plus a per-trunk charge (WATS trunks). Still others are billed at a flat rate no matter how much (or little) they are used (tie trunks).

Purchase of trunk facilities and the rules for selecting them are principal tools for lowering the long distance bill. The choice between how many fixed-rate-billed facilities to purchase and how many usage-sensitive-billed facilities is predicated on how the system will direct

TABLE 5.3   Facility Administration Parameters

Trunk types
ACD
ANI
Call park
Call vectoring
Centralized attendant
Code calling
Conference
Data call setup
DID
DOD
DS-1 interface
Host computer interface
ISDN-PRI
Loudspeaker paging
Modem pooling
Radio paging
Remote access

traffic between these facilities. These rules are contained in parameters used by features such as Automatic Alternating Routing, Automatic Route Selection, Code Restriction, Facility Restriction Levels, and Route Advance.

To achieve expected economies, the rules programmed into the PBX must accurately reflect the assumptions used to decide quantities of various facilities to purchase. Errors made here can adversely affect the security of access control to long distance services and business computers resulting in substantial revenue losses, costly interruptions of the business activities, or compromise of proprietary information.

Traffic Management

Traffic Management (traffic administration) is a function that enables the end user to gather information on call volumes handled by the PBX and its server groups and analyze that data. A server group is any group of facilities or agents that are selected from a pool to provide a common service. Trunks in a trunk group, ACD agents in an ACD split, and attendants are all examples of server groups. Many PBX components are traffic-sensitive. That is, they are provided in sufficient quantities to assure a specific grade of service (blocking probability) or average service delay. For example, a PBX responds to a request for service by connecting a DTMF receiver to the time slot assigned to the requesting station. The DTMF receiver provides dial tone and captures the address signal. The quantity of DTMF receivers needed to assure satisfactory service (that is, a minimal dial tone delay) is traffic-sensitive.

A PBX populated with heavy users would require more DTMF receivers than one with light users for the same quality of service.

Traffic levels within a PBX are dynamic. At times the traffic may be heavy (busy hour, busy day, busy week); at other times, it may be relatively light. Server group grade of service or average delay is usually specified for the busy hour. Selection of group sizes to achieve these service standards or objectives is called *engineering*. A group is said to be "engineered for p0.01 blocking," meaning that one out of every one hundred attempts to obtain service during the busy hour will find all servers busy and be blocked. (The call isn't necessarily actually blocked. It may overflow to another trunk group.) To verify service quality during high traffic periods, the traffic management capability allows the end user to specify that specific traffic measurements be captured on specific components at stipulated times. This data is then captured and analyzed by the user to determine if the system is properly configured.

Traffic Management is used to monitor and manage all shared resources within the PBX. This includes groups of people serving a common function such as attendants and Automatic Call Distributor (ACD) splits. Shared resources are those that are used by many PBX users at different times, usually, but not always, on a first come first served or First In First Out (FIFO) basis. Common examples of shared resources are trunks, DTMF receivers, network paths, and the internal processor.

The first step in traffic management is to decide what the policies and priorities will be for use of shared resources. Who will be allowed to use the facilities, for what purposes, at what priority, and at what times of day? This can be illustrated in access to long distance services. Let's assume that a company has two major locations, one in Boston and another in Denver. Let's also assume that the people in Denver need to talk to people in the Boston location regularly, but do not need to talk to people elsewhere in Massachusetts. After doing a traffic study, the company concludes that they can economically lease eight tie trunks between Denver and Boston. All traffic that can't be carried on the eight tie trunks will overflow to the CO trunks and be carried by the public network. To enforce these decisions, the company uses Code Restriction to allow only calls to the Office Code of their location in Boston within the 617 (Boston) area code, and none to other Massachusetts area codes or office codes.

Because of the cost of public network calls, they restrict access to these facilities to a few management personnel. All others who find the tie trunks busy must wait for one to become idle. If the PBX has Callback Queueing, it can call users from a priority queue when the trunks become free.

Having established these policies and sized the trunk groups according to the earlier traffic study, the question becomes "How's it working out?" Things change as time goes on. Project interactions change, locations grow or decline, people change, and projects come and go. It is not a static one-time task to size facility groups and assign permissions. The data to answer these questions comes from Traffic Management and Cost Management. Traffic Management provides the measurement and reports on how the existing facilities are being used. Cost Management (to be discussed later) provides the detailed data on who is using them. When decisions are made between purchasing tie trunks (a fixed cost per trunk) and using the public network (via CO trunks), the question of how many trunks to purchase is one that says the last trunk in the group has to carry enough traffic to be more economical than carrying that same traffic on the public network. This is sometimes called the "Economic CCS." (CCS is an abbreviation for 100 call seconds per hour. A single trunk can carry a maximum of 36 CCS since there are 3600 seconds in an hour.) The only way to know how things are actually working is to make measurements and compare results with expectations.

Similarly, for a group of agents (say attendants), initial sizing of the attendant group was based on certain assumptions or measurements of how much traffic the group would handle. If there are too few attendants, the customers will experience long delays in having their calls answered (calls to busy groups of agents are typically queued rather than blocked). If there are too many agents, the customers will be happy, but the cost of paying the staff will be too high. Sizing agent groups is more dynamic than sizing facilities, because not only can you adjust the group size, but the staffing levels can be changed throughout the day to fit demand.

Traffic Management provides the data and tools to manage these facilities and groups accurately according to customer-determined performance criteria. Widely used statistical models of telephone traffic for all these situations and many others are available to allow the customer to choose the performance that best meets their business goals at the lowest cost. For systems that reject service bids (block) when all facilities are busy, the measure is grade of service (the probability that a random call will find all facilities busy and thus be blocked).

For systems that queue service bids (making them wait in line) when all facilities or agents are busy, the measures are *probability of delay, average delay,* and *average delay of delayed calls.* The difference between average delay and average delay of delayed calls is average delay in total time for bids in queue divided by the total bids, whether delayed or not, and average delay of delayed calls is divided by the number of bids that were actually delayed. From a customer satisfaction point of view, there can be quite a difference. If there are 100 calls to an attendant and one of them gets queued for 300 seconds before being answered, that customer will be pretty unhappy. The average delay is 3 seconds (sounds pretty good) but the average delay of delayed calls is 300 seconds! How each of these numbers is used is a business decision potentially affecting revenue and cost. The important point is that these choices can be made on a business basis and Traffic Management provides the tools and data to enforce the decision.

Traffic Management also includes planning telecommunications facilities for expected growth or downsizing. Either of these events would change the expected load on the facilities from what is currently measured, requiring a facility group size change. Here, the same procedures are used as for sizing operational facilities or groups, except the traffic data are modified to reflect the changes expected from growth or downsizing. Planning for such changes and placing the necessary facility change work orders in time to support the projected changes is part of Traffic Management. The extent to which it is done can affect both cost and user satisfaction.

Cost Management

There are many costs incurred in operating a telecommunications system. These include investment costs for purchasing the PBX; building wiring and telephone sets; operational costs for attendants, system administrators, and installation and repair personnel; building space, power and air conditioning; and lease and service charges for long distance calls and public network access.

Cost Management is the determination of how the costs of telecommunications should be allocated among the PBX users. A principal reason for allocating costs to users is to place management and accountability for those costs where management controls can be implemented effectively. The first requirement for effective Cost Management is the establishment of policies for who can use telecommunications facilities and how telecommunications costs will be allocated. Will the operational and investment costs be allocated uniformly on a per-user basis or will they be usage-sensitive (based on the total number of calls) or some mix of number of users and system usage? Will long distance charges be billed to the user (or their organizations) or will this be a corporate expense? How will the costs for leased facilities (tie trunks) be distributed? Will they be billed to those users that actually used the trunks, or all users whose calls accessed routing algorithms that include the trunk group?

Cost Management is an adjunct management feature responsible for implementing the cost allocation policies. It requires the collection and analysis of data to determine the system usage and equipment assigned to each individual for purposes of "fairly" distributing the costs of operating the system to the user community. The system usage data are collected as the calls are made in internal PBX records called Message Detail Recording records. Records of the equipment assigned to each user are maintained in a separate inventory database. Sometimes these inventory databases are part of the PBX administrative database. These data are combined in off-line data reduction programs to calculate the telecommunications bill. These bills may then be aggregated by user, organization, project, or any other parameter the customer decides to allow responsible individuals to identify and control their costs and pay their fair share of the telecommunications bill.

A stored program PBX must retain a record of every billable call it handles to provide Cost Management with the necessary usage data. In these records, the PBX keeps track of the originating port (translatable to originating terminal or trunk), the destination party (translatable to destination terminal, trunk or ARS/AAR routing pattern), the called number, the time and date, the facilities (trunk) used, an optional account code (for billing by project), calling party access privileges (Facility Restriction Level, Authorization Code, etc.), and the duration of the connection. These Message Detail Recording (MDR) records are queued within the PBX until extracted by an adjunct system. These adjunct systems are often designed to extract and process MDR records from multiple systems and aggregate data from across a private network. The adjunct system can then use algorithms to rate the call and prepare detailed bills or similar billing-type records by individual, department, etc.

Processing of MDR records to determine the telecommunications bill is usually done off-line on a periodic basis (say monthly). It is important to provide timely bills so that these costs can be managed by those responsible, but aggregation of data by the system is necessary to make the data easily understandable and differentiate calling trends from occasional calls. A long list of individual calls is difficult to comprehend. Off-line processing (not the active call processor) is necessary to minimize the effect on call processing capacity from MDR data reduction work that has much less stringent real-time response requirements than other PBX operations such as call processing.

Cost Management adjuncts can be designed to extract administration data from the system or system management database. These data provide details about the users' terminals and ancillary equipment. This reduces the possibility of using erroneous data for billing by assuring that a common database is used for call processing and billing. Similarly, gathering user usage data through the traffic management system can allow a cost management adjunct to allocate PBX common costs to users and departments. Traffic Management data provides a system view of how trunk groups are being selected by internal call routing algorithms (AAR, ARS, etc.). These routing algorithms are associated with specific Area Codes, Office Codes, and private Network Node Identifiers (NNXs), and determine which called number destinations are generating leased facility use. Traffic Management data are sometimes used to allocate the fixed costs for these leased facilities.

System Management systems of the future

Vendors are already making significant progress in improving system management tools. Modern screen- and task-oriented administration software have substantially reduced the work required to perform a move or change, and improved the reliability with which changes can be entered. Off-line databases are uploaded from the PBX to an administration system (to assure data consistency with the switch data) and changes made according to accumulated work orders. These can then be scheduled to occur as appropriate. They might be implemented at midnight Saturday to correspond to a large reorganization to become effective over the weekend or immediately after entering the data.

Any attempts to improve current administration systems probably will be primarily directed at the terminal and station administration problem. This represents about 95 percent of all administration activity, and improvements in administration must either make this function more reliable or reduce the labor required. The steps in terminal and station administration are:

- Logically associate the user with the terminal. This means choosing a terminal type, assigning an extension number, assigning service access privileges, and assigning the user name to the extension number for directory.

- Logically associating the terminal with the PBX software. This means assigning terminal buttons and displays to features and extensions.

- Physically associating the terminal with the PBX. This means deciding and assigning where the terminal shall be (or is) wired into the PBX.

Most companies will have a few (say, six to ten) classes of employees that will be given similar terminals and feature access privileges. This means that most of the work of logically associating the user with the terminal, and the terminal with the PBX, can be done using a few standard assignment templates. These templates can be stored in the administration system and recalled to begin adding a new user. Adding the extension number and the user name will be unique. Other ad hoc changes to the template could be made to reflect individual needs and preferences.

Physically associating the terminal with the PBX is always unique. This is usually done by mapping cable runs from the physical location back through the cross-connect field to the PBX port circuit. Much of these data are manually collected and maintained, especially the cable pair assignment data; therefore, it is error prone.

An alternative is to enter all the information on a user via System Administration except the physical port where the terminal is connected. When the user goes to their new work location, they plug their terminal in, get dial tone, dial a security code followed by their extension number. The system recognizes the security code as saying this call is to be a physical-to-logical mapping and uses the extension number to identify the appropriate mapping.

Administrative controls would be needed to control the time window where a user would be permitted to move their terminal. This would prevent the uncontrolled migration of terminals or extension numbers around the premises with the accompanying inventory control and long distance access control problems. Computer-controlled hardware has been designed to identify itself to software so that software will use the available physical configuration for some time. This is an application of that principle.

This scheme also would facilitate moves since most moves would not require a new terminal, different feature or facility access privileges, or different terminal button assignments. Just have the user move the terminal, plug in, and identify themselves. The system would do the rest. This has the following advantages in facilitating moves:

- Reduces the labor required to make a move. The system administrator only has to open the time window to allow the move. Most of the system administration work for operational systems is user moves. This can significantly reduce the work required.

- The timing of the move implementation is under the control of the user, assuring coordination with other activities associated with the move. (Did the furniture really get moved?)

- The reliability of mapping the terminal to the correct port is assured. Cable record accuracy and the labor of ringing out the circuit ahead of time is no longer necessary. If there is some connection to the correct type of port circuit, the system will find it.

- Users could be assigned particular terminals instead of locations. This places accountability for terminal inventory with the individual having physical control of it as with other office equipment.

The idea clearly needs discussion and development. It raises new questions regarding system security and access control. User reaction to taking some responsibility for doing their own move needs to be evaluated. I believe that the added user control of the move process will be well received. An unanswered question is, "What new and unexpected applications might be made of this functionality?" For example, will executives want to move their terminals to, say, a conference room to facilitate teleconferences with convenient access to familiar features and speed calling lists? Will users want to temporarily move terminals off premises (say to their home)? This would be entirely possible using ISDN. What new security problems for service access might this added flexibility create?

Vendor upgrade policies

An issue that deserves the users' attention when a PBX purchase decision is being made is the vendor's position on upgrades. End users want upgrades designed so that wholesale changes to the administration parameters are not needed. Often, this is possible with a well-designed change. However, if the system generic program or operating system is changed, it may be impossible to avoid such database modifications. Changes in the generic or the operating system often result in changes to the basic design of the database, especially in how data is stored. This happens because data representations often use only a few bits of a memory word rather than the whole word. This is done to reduce the amount of memory required to store all the data. When the new generic or operating system is introduced, the values in the fields of the old program must be read, interpreted, and converted to the new format.

When this occurs, vendors have several choices. They may choose to write software that can be run on-site that translates the administrative database to the appropriate new format. They may offer to copy the administrative database before the upgrade and enter the revised database information at the time of the upgrade. This solution creates an additional problem in that there is always a time lag between when the database is copied and when the new program is cut into service. Administrative change activity continues during this time. When the cut occurs, the changes made during the interim period must be reentered (often manually), which is an error-prone process. Because of these difficulties, when this change procedure is to be used, every effort must be made to reduce the interval between when the database is copied for translation and when the new software is cut into service. Additionally, change activity during this transition period should be limited to only those changes that are essential.

The vendor may decide that work on the administrative database is an end-user responsibility. They may provide the end user with the information on modifying the database and coordinate the upgrade so that the user simultaneously updates the database. Or they may choose to leave the problem in the lap of the end user to be resolved after the upgrade is installed. The problem with the latter position is that the system will not function properly until the administration database is updated. Considering the difficulties and risks described above, any process that minimizes vendor participation in a major upgrade should be examined very carefully for feasibility.

Maintenance

System Maintenance includes all functions required to detect and analyze problems and take appropriate corrective action to restore service. From a practical point of view, installation of new facilities, terminals, and services are also the functional responsibility of the maintenance organization. System Maintenance, ideally performed, would keep the system fully operational twenty-four hours per day, three hundred sixty-five days per year. Clearly, this is not an achievable goal with today's technology when the associated costs are considered. We will explore the issues and possibilities in this section to provide a framework for choosing among available options. Fig. 5.2 illustrates that System Maintenance can be subdivided into four major categories.

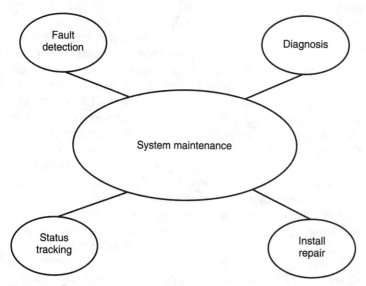

Figure 5.2   System maintenance functions.

The systems involved in the maintenance process must be capable of performing the functions listed for problems that arise in system hardware, software, and firmware. Considering the complex architecture of the modern PBX and the need for highly reliable communications, the design of reliable maintenance systems is critical.

Fault detection

Recognition of conditions that indicate actual failure or potential failure is called *fault detection*. This includes all the ways faults are identified and reported. Ideal fault detection and repair finds and clears faults before users are aware that a fault has occurred. While this objective is not completely realized, most faults in a modern PBX can be detected and cleared without user-perceived service degradation. For example, a trunk may become noisy or fail in its supervisory signaling sequence. Trunk fault detection hardware and software detect this condition, take the trunk out of service, and generate an alarm to get the faulty trunk repaired. Here, a user may have tried to use the trunk and experienced a call failure. Simply retrying the call will cause a good trunk to be selected and the call will complete. Though the user has been affected, they probably will not perceive it as such.

Fault reporting can be a result of a built-in test or a service complaint from a user. The modern PBX has a hierarchy of self-tests built into the system. Tests are generally designed to detect problems at the lowest level possible. This provides for the fastest isolation of a fault to the failed unit, and distributes the workload for fault detection to many processors and circuits, resulting in a smaller impact on call processing capacity and a faster fault detection time. These tests, together with user complaints, are the source of all failure alerts to repair personnel.

The first level of tests is built into microprocessors or circuits on individual circuit boards. For example, a T1 interface board contains tests of the transmission quality being experienced on the T1 link. Quality measures such as errored seconds and frame slips are continuously monitored. Should these exceed predetermined limits, an alarm is generated to the software. Similarly, packet-switching interface boards monitor packet transmission performance parameters such as number of retransmissions and CRC error corrections. Error rates are thresholded to determine "acceptable" or "unacceptable" performance.

The next level in the hierarchy of testing is event-based software tests. These are tests performed by the call processor each time a specified event occurs, usually the successful or unsuccessful completion of a call processing task. An example of this kind of testing is looking for extremely short or long holding times for calls on a trunk. Short holding times can signal a *killer trunk* (one where the supervisory interface is failing to hold the call up). Long holding can signal a *hung trunk* (one where the supervisory signaling fails to drop the trunk connection when the parties hang up). The thresholds for *short* and *long* are often administrable to accommodate site-specific use practices. If a company is using trunks to access computers, long might be several hours. Similarly, if a location is using dial-up trunk connections for credit card verification, short might mean a couple of seconds. Another example of this type of test is processor sanity checking. Often there is a task in the call processors worklist that must receive processor service within a specified period, which is measured by an internal clock. Failure of this task to receive service within the specified time is assumed to be due to processor failure or "insanity." Response to this type of error may be raising a major alarm, restarting the processor, switching to a redundant processor, or any combination of these.

The next level of testing is periodic software tests. These tests are performed by the call processor periodically on a scheduled and time-available basis. These tests try standard operations whose expected results are known, and verify that the expected result is obtained. They also do audits of the call record database (the database of information on active calls), looking for inconsistencies and correcting them if possible and logging test results. For example, they may try to seize a particular idle touchtone® receiver. If successful, it may try another. If it fails, an alarm may be generated and an error log entry made indicating the defective receiver and the particular test that failed.

These tests contend with call processing tasks for processor real time, yet usually they do not have the tight real-time response requirements of call processing tasks. Therefore, they are scheduled at a lower priority than call processing on a time-available basis *most of the time.* Execution of these tests becomes urgent if they are never being executed. If the call processor is too busy to get to any of these tests for an extended period (say several seconds), this could indicate problems with call processing. Now, the priority of these tests is raised above call processing so they can appraise the "health" of the processing system. While execution of these tests is not usually urgent, the rate at which they get executed over a period of time affects the mean fault detection time. If a fault occurs at some time that would be detected by one of these programs if it were run, but it doesn't run for, say, an hour, system availability (availability will be discussed later in this chapter) could be adversely affected. Design of the maintenance system should include consideration of how long the permissible *cycle time* (the time between subsequent running of the same test on the same equipment) is, and assuring that processing capacity under heavy call processing load is such that the needed reserve capacity for periodic testing is still available.

The next level of testing is *demand testing*. Built-in tests can be performed both periodically or upon demand by a technician or a separate controlling maintenance facility. Test results may be reported immediately via an on-line display or alarm or they may be logged in an error log for thresholding to determine if they exceed some preestablished level. Usually when the number of errors exceeds a threshold, a minor alarm is raised. Problem isolation can begin with additional information on the nature of the error from the error log being used to initiate appropriate additional tests to further isolate and identify the problem. Often these tests are run from a remote site with an objective of clearing the problem remotely or dispatching technicians with the correct replacement parts to clear the problem.

Remote problem diagnosis reduces the cost of maintenance by using the maintenance forces more efficiently. Technicians staffing remote maintenance terminals are often selected because of their demonstrated superior skill. Working at a fixed location on problems reported from many systems gives them more experience in solving problems, and they don't spend much time traveling from one location to another. Many reported problems can be cleared remotely, using maintenance and administration software, thereby avoiding dispatching any maintenance personnel. For example, a problem might be created by entering incorrect administration data.

Some reported problems are intermittent with a sufficiently low occurrence rate that maintenance personnel cannot really clear them until they become more frequent. These are often reported "no trouble found." "No trouble found" rates have been known to run as high as 50 percent of all trouble reports. Reducing the dispatch rate and assuring that the technician will have the problem evaluated and have the correct spares greatly reduce maintenance costs.

Maintenance performed remotely usually accesses the maintenance software through a dial-up port. These remote maintenance ports are a point of system vulnerability for intruders seeking to use or disrupt telecommunications facilities for profit or recreation. Consequently, these ports usually have some form of security protection such as passwords to make clandestine penetration more difficult. Adequate security protection for these ports is essential! This will be discussed more thoroughly in Chap. 6.

Installation and Repair

Installation and repair includes all activity to clear problems and to install or move hardware. These functions are usually grouped together because common technicians perform both tasks. Combining these functions simplifies the management of clearing problems created by improper installation techniques, and benefits from skill development common to both tasks.

A primary consideration in the cost of maintenance and installation forces is their level of training and expertise. Well-trained and experienced technicians are much more efficient at performing their tasks. Consequently, it is often less expensive for companies with just a few lines or PBXs to contract maintenance to organizations maintaining many other PBXs of the same type as theirs, rather than assume responsibility for training and management of a maintenance organization.

Repair activities include system, software, hardware, facility, and interface-type problems. System problems include such things as administration, system environment, and utilities. Software problems include software bugs (some bugs do find their way into the field) and patch installation. Hardware problems are all problems that require replacing hardware to clear. Interface problems are those associated with connecting the PBX to equipment or facilities such as trunks, terminals, adjunct processors, and computers. Installation includes adding or removing terminals, trunks, adjunct processors, or any other equipment or software to the telecommunications system.

Besides isolating and clearing problems and installing new equipment, this task also includes management of the spares inventory. This includes both spare parts and cable pair assignments with a record of unused pairs in the cables. The cost of maintenance can be greatly affected by how well this job is done. Cost of maintenance personnel can be greatly reduced by maintaining on-site spares for the most commonly replaced circuit packs and equipment (usually terminals and line circuit cards). On-site spares can greatly reduce the occurrence of finding a problem and not having the necessary spares to replace the defective component, thus requiring a second dispatch. Maintenance technicians are well aware of this and in the days of the old Bell System often kept private repair stock in each PBX. Modern inventory control systems make this much more difficult when it's not the official policy. Similarly, when a new terminal is to be installed, good cable records are invaluable in quickly finding cable pair to make the necessary connections.

Status Tracking

Managing repair forces and identifying and clearing intermittent problems requires tracking of the status of all open problems and work orders. It also includes maintaining records on problem histories for equipment and facilities to identify any long-term trends that may suggest that root problems are not being cleared such as intermittently defective trunks in the central office. A computer-based Trouble Ticket system that includes status of all open trouble reports is valuable in answering user questions about progress in clearing problems and preventing reported troubles from "falling down the crack." These records are also used to determine repair force performance and adjust vendor bills where the contract includes quality performance clauses.

Maintenance system architecture

Modern PBXs rely on effective software systems with supplemental hardware to perform much of system maintenance. Software systems gather indicators of trouble from hardware systems, often from on-board processors. These systems also log each incident of trouble encountered in setting up calls. They also log incidents of logical errors encountered in call processing software. Software systems will periodically perform audits, searching for anomalies in the system, and will either correct the condition or log it. The software systems will often have diagnostic subsystems that will do a thorough analysis of the condition and determine the problem cause. Finally, the systems will ascertain the potential or actual magnitude of the effect of the problem on system performance and will turn on the appropriate alarm (usually major or minor). Sometimes, these maintenance software systems will busy out or switch failed components. For example, maintenance software can initiate a switch to a redundant processor.

There are several maintenance interfaces to the world outside the PBX cabinet. While they appear distinct, they are in fact integrated. One is the alarm interface. The alarm interface indicates the status of the major and minor alarm indicators. This alarm status information can be displayed locally or arranged to be transmitted to a remote location such as a remote maintenance facility or a network management system.

Most PBXs use a major/minor alarm scheme. Generally, a major alarm is initiated when a significant service-affecting failure has occurred or the system determines a major fault is imminent. The major alarm is designed to prompt an immediate response by maintenance forces. Besides catastrophic failure, there is a wide variation among vendors on what constitutes a major alarm condition. Typically, maintenance personnel have to retrieve additional information from the PBX to determine the cause of an alarm. This information is usually obtained from the error log and possibly from remote testing.

"All other" conditions are indicated by minor alarms. This can be a bothersome category in improperly designed systems. There can be so many minor anomalies in a software-driven system that the indication of a minor alarm may be continuous. One solution is to have programmable thresholds for a variety of problems that prompt error log entries. Minor alarms are not initiated until one of these thresholds is exceeded.

The second interface is the system of error logs and error message indicators. This is where the maintenance system records the information it has uncovered and diagnosed concerning PBX performance. This is the database that maintenance software uses to initiate an alarm indication. As noted previously, thresholds can be established that set the level below which an alarm is not generated.

The user interface systems used with PBX maintenance systems usually are dumb terminals or product-specific devices for small systems. These are simply designed to be as low in cost as possible and work directly with maintenance software built into the PBX. They simply provide access to the software built into the PBX. Larger PBXs have moved to software-based maintenance tools, placing more intelligence in the terminal for screen-oriented presentations, error checking, and data analysis. These tools are also product-specific. They are designed to complement the PBX hardware and software architecture. Communications between the maintenance tools or devices and the PBX use proprietary protocols.

Artificial intelligence is being incorporated into adjunct maintenance systems. These AI-based systems can receive an alarm transmission from a PBX, log an entry, analyze the data received, and initiate a trouble ticket. Upon receiving an alarm, the system calls the maintenance port of the PBX and interrogates the system. It will cause additional diagnostics to be run if the condition is ambiguous. Upon receiving the error log data, plus the results from any requested demand tests, the system determines the cause of the trouble and the required maintenance action to clear it. Since these are AI systems, the analysis is not restricted to analysis provided by the vendor, but it is capable of "learning" from its own "experience." The main objective of the AI system is to provide the maintenance personnel with an exact diagnosis of the problem cause before they are dispatched. Alternatively, the AI-based system can cause an alarm to be retired if it ascertains that the condition is spurious or it is a nuisance alarm.

Administration and Maintenance in Network Management

The PBX maintenance system must fit into a larger maintenance system architecture if it is to be used in a network. PBXs often are elements in communications networks handling both voice and data traffic for large corporations. Many of these networks are integrated with a common network management system. To fit into such an integrated system, the PBX or its maintenance controller must support the larger network management protocols.

Though the PBX administration and maintenance systems are product-specific and their communications protocols are proprietary, still they fit within the industry architecture for network management systems. Network management systems grew out of the need to maintain reliable communications across an entire enterprise, regardless of whether the communication was data or voice. As the technologies for each evolve, they become more similar, therefore the problems and functions needed to maintain both are similar. It was inevitable that they would evolve into a common management system. Network management system architecture has evolved to a three-tiered structure. The architecture is illustrated in Fig. 5.3.

The first tier of the network management architecture contains the components of the network. Devices such as hosts, multiplexers, transport systems, LANs, and PBXs are network components. These are often called *network elements*. The network element is the focus of the network management system and is the "managed object" within the hierarchy. The network management model with subdivision of components into manageable objects is called object-oriented network management.

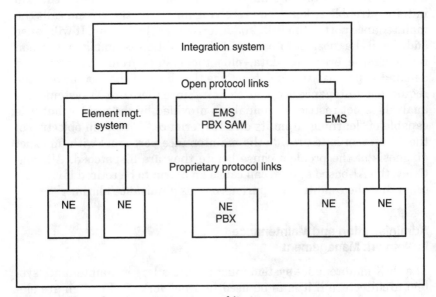

Figure 5.3 Network management system architecture.

The first tier can be further subdivided into coded objects. A PBX would be a major category of coding. But a PBX is composed of components that can be administered or maintained (i.e., managed). Thus, the PBX can be subdivided (subcoded) into different types of manageable circuit packs. Finally, since individual ports of circuit packs can be individually managed, a subdivision is made at the port level.

The second tier of the network management consists of those devices that are used to manage the network elements. These element management systems are what we have described in this chapter. PBX administration and maintenance systems are element management systems. Element management systems are typically developed by the element development organization (e.g., PBX management systems are developed by the PBX developers). This is because the interfaces to the element maintenance and administration systems are usually proprietary and the detailed knowledge and understanding of the element management design is only comprehended by the element developers. Also, this facilitates timely update to the element management system whenever new functionality is added in ongoing product evolution. The network management hierarchy assumes that element management systems are product-specific and that the communication with network elements is proprietary.

The third tier of the hierarchy consists of the systems that integrate the communications and information flow with the tier 2 element management systems. The interface between tiers 2 and 3 utilizes industry-accepted open protocols and standards. Currently, the International Standards Organization is developing a network management protocol and set of standards based on their Open Standards Interconnection (OSI) network model. Other industry leaders are working to get a Simplified Network Management Protocol (SNMP) that supports the popular Transmission Control Protocol/Internetworking Protocol (TCP/IP) networking standard.

Reliability

Reliability is a term that is very often misunderstood and very often misused. Reliability is defined as the extent to which an experiment, test, or measuring procedure yields the same results on repeated trials.

What is it that customers want? Through user groups and other forums, end users have expressed the position that they want the PBX to be capable of responding completely and promptly to their requests for service. That is nearly in keeping with the definition of reliability. This doesn't cover all concerns regarding failures, however. If a fully redundant component fails, the system may "respond completely and promptly" to a service request. Nevertheless, a dispatch of repair personnel will be required to repair the failed component. What this suggests is that whether a failure is deemed to have occurred or not is dependent on your point of view. If you are a user, you just want the system to work. If you are a repair organization, there is also a question of whether you had to take a repair action.

A situation that causes significant disruption to the PBX's capability to respond to service requests is called a *major outage.* Users have made their views clear. They call a major outage the loss of one or more of the following services:

- the ability to originate outward calls
- the ability to originate calls to the public-switched telephone network
- the ability of the system to process incoming calls
- the ability of the system to process intrasystem calls
- the ability to access private networks

Some businesses consider the loss of some services provided by adjunct processors to be an outage. The service of most concern is CDR that provides the data for Cost Management. PBXs are designed to be able to store many CDR call records while waiting for an external polling device to retrieve them. Storage capacity is limited; therefore, the length of time that the polling system can be down without overflowing the storage capacity of the PBX is limited and a function of PBX traffic. When the record storage capacity of the PBX is reached, some businesses elect to block outgoing calls until the stored records are retrieved. This creates a major outage. The outage results from a business decision to block calls if CDR data cannot be collected.

Measuring Failure Rates

The commonly used measure of system outages is the Mean Time Between Outages (MTBO). The major determinants in MTBO are the size of the system and the degree of redundancy. Typical MTBOs for PBXs range from several months to tens of years. Vendors customarily use MTBO to advertise their system reliability because it is often measured in years.

Often, PBX vendors will advertise data relating to how long a time their product is up and ready for use. Does MTBO adequately measure system reliability? Certainly it expresses the readiness of the system to respond completely and promptly to requests for service *by all users.* But how well is the individual user served?

Failures in the system caused by faults in components, the end of life of components, or person-made errors can affect the entire PBX community (an outage), part of the community, or a single user. End users prefer that failures be measured from the single-user-failure perspective. The generally accepted description of a single-user failure is any event that prevents a single user from receiving or placing a call for 15 seconds. An example of a single-user failure that would not affect MTBO is the failure of a line circuit card. If you are the user connected to the line circuit card that failed, your telephone doesn't work! Most failures in the system are of the single-user type and all users will experience more frequent failures than suggested by the MTBO. This leads to system users having a much worse impression of system reliability than the published MTBO numbers and becoming cynical regarding vendor performance claims.

All service delays are not the result of system failures. Excluded from consideration are services that are deliberately delayed or where shared-usage queueing is employed. For example, if a user is placing a long distance call by accessing a small flat-rate-billed trunk group and is queued for access to the first available trunk, no failure is tallied. Queueing of service to shared-use facilities is a common technique for providing all users with service at a lower cost. It is most often used with fixed-rate trunk groups (tie trunks and sometimes WATS trunks) and groups of agents such as attendants or ACD splits. It is a technique for getting greater average usage (thus lower average cost per access) with minimum inconvenience for the users.

The accepted measure of failure rate is Mean Time Between Failures (MTBF). Single-user MTBF for typical PBXs is considerably shorter than MTBO and usually ranges from weeks to months. Some vendors will use the terminology MTBF but use MTBO criteria. Since there is no industry standard for these terms, it is prudent to ask the vendor to define what terms mean as used in their product descriptions.

Measuring Downtime

Data that measures the failure rate of a PBX is only part of the equation to evaluate system reliability. The other necessary information is the repair time. Even a concept like *Downtime* is not as obvious as it might seem. When does the downtime clock start? There are three candidates:

- when the failure actually occurs
- when the failure is detected
- when the failure is reported

A person might think that the clock should begin when the failure occurs. This has a problem analogous to the classic question of "If a tree falls in the woods and no one or thing is there to hear it, does it make any sound?" If a failure occurs and it is not noticed or detected, what does it mean that a failure has occurred, and who cares? If this principle is carried too far, automatic testing would be suspended because it only speeds the time when the failure is detected. The real world problem with this starting time is that it is ambiguous because it wasn't detected at its onset.

The time the failure was detected is the first really logical candidate for time of failure. Users would like to use this time because it is the first point at which their service might be affected. Even this time has some ambiguity. For failures detected by automatic test faults exceeding a threshold, was the failure detected when the first test fault occurred or when the threshold was exceeded? These questions can usually be settled through negotiation.

The part of the telecommunications industry that services PBXs generally agrees that repair time is the elapsed time from when a failure is reported to the maintenance (service) organization until service is restored. The reason is obvious. They want to respond to users who want a bottom-line statement on how good service will be (service objectives often have penalty clauses that come into effect if the service objective is not met). Service vendors do not want to be responsible for not responding when the failure has been detected but they have not been notified. Reconciling the user's view and the service vendor's view involves two issues. First, does the end user report the problem promptly? Second, if the system detects a potential fault that does not result in a single-user failure and initiates an alarm, but maintenance action is delayed, does the entire elapsed time constitute repair time? No universally accepted position is available today.

The accepted measure of repair time is Mean Time To Repair (MTTR). End users would like the vendor's MTTR to be about 2–4 hours for smaller systems and 2–8 hours for large systems. The factors that can affect MTTR if they use vendor or third-party service are:

- Whether the user is under a maintenance contract or an occasional-use agreement. Occasional-use agreements are often called time and material (T&M) agreements. Nearly all vendors and third-party offerers give priority to their maintenance contract customers. They quote fixed response times to these customers and "best effort" response to T&M users.

- Technician training and skill level.

- Distance between the PBX location and where the technicians are located. Major vendors and third-party maintenance organizations can provide dedicated, on-site technicians to end users.

- The design of the PBX maintenance system. If the PBX has a good diagnostic and alarm system, trouble diagnosis can be greatly improved.

- Customer-owned and -deployed maintenance stocks can drastically affect repair time. If the technician must order replacement parts from stock, the repair time can run into days.

PBX system outage, failure rates, and repair times always have been of primary interest to designers, vendors, and service contractors. Outage and failure rates are determined by the design of the system and the environment in which it must operate. Repair time most often is affected by the quality of the repair force and, to a lesser degree, by the design of the maintenance facilities provided. They also are useful in determining life cycle costs of systems, and thus should be of interest to end users.

The use of redundancy to improve reliability often is misunderstood. Redundancy adds more components to the system and increases system complexity. The overall failure rate (the MTBF) theoretically, and practically, will increase.

What redundancy does is to reduce the repair time (the MTTR). For example, automatic system tests monitor the health and performance of critical system components such as the central processor. When a fault is detected, a standby processor is switched into service. Depending on how tightly the two processors are coupled, the switch can take anywhere from a fraction of a second to several minutes. The loosest coupling has a spare processor ready to be turned on and put into service. The tightest coupling has the spare processor performing all the call processing operations in parallel with the active processor to assure that it has a completely up-to-date active call database. This is called *shadowing*. Using processor shadowing, the standby processor is ready to take over call processing on a moment's notice and continue operation virtually unnoticed by the user community. It is possible that some calls in the process of being set up will not be handled correctly because of the failure causing the processor switch. The general user community, however, perceives a seamless operation of the telephone system.

Another simple application of redundancy is provisioning trunk groups with more trunks than required to carry the offered traffic at specified service levels. If a trunk in the trunk group fails, automatic tests detect the failure and take the trunk out of service. Since the trunk group has more trunks than required for the specified grade of service, service to the user community continues to meet these service objectives. This solution is expensive for flat-rate trunks such as tie trunks.

Using redundancy, service is restored quickly, perhaps in zero time. In a well-designed system, the intent of system redundancy is to reduce the MTTR dramatically while modestly increasing the MTBF. Thus, the availability is significantly improved. It should be recognized that strategically placed redundant components can decrease the outage rate (the MTBO) because major failures no longer cause service outages.

System availability

If reliability is a consideration of interest to end users, then the measure of system availability is a metric of primary concern. Availability is related to downtime and depends on failure (or outage) rate and repair time. It is defined by the equation:

$$A = \frac{FR}{FR + RT} \times 100$$

where   A = percent availability
        FR = failure rate (MTBO or MTBF)
        RT = repair time (MTTR)

Availability is a measure of the probability that the system will operate when the user accesses it. Usually, it is measured on an outage basis but can be measured on a single-user basis.

Let's assume a PBX experiences a single outage in a year that takes eight hours to repair. The availability is:

$$A = \frac{365 * 24}{365 * 24 + 8} \times 100$$

$$= 99.91\%$$

It seems that in 10,000 attempts to use the system, the body of users would be successful 9991 times. What it really says is that in 10,000 attempts, on the average 9 would fail due to a major system outage. Notice the word "average." It hides many assumptions—for example, that the failure occurs during an average traffic period. If business users place three calls per user during the normal eight-hour workday and none the other sixteen hours, the average traffic period has one call per user. No time of day actually has traffic of one call per user! The point is that these are useful comparisons to make between systems, but the actual values, if taken literally, need thoughtful analysis to make meaningful comparisons with actual results.

If a failure affecting a single user occurs once a week and requires a repair time of four hours, the single user availability is:

$$A(su) = \frac{7 * 24}{7 * 24 + 4} \times 100$$

$$= 95.45\%$$

Single users will be successful 9545 times in 10,000 attempts with the same analytical caveats as discussed above for system outages. Interestingly, because of the larger numbers, the single-user-failure experience is more likely to approximate the calculated values than the major outage failure rate. This is the Central Limit Theorem from statistics in operation.

Dispatch Rate

Another performance parameter of interest to organizations doing maintenance is *Dispatch Rate.* This is the rate at which it is necessary to dispatch a technician to clear problems. The dispatch rate is typically much lower than the failure rate because many problems can be cleared using remote maintenance; others are classified "no trouble found." Dispatch rate will be affected by the quality of the remote maintenance and testing software, and the skill of the remote maintenance staff. For example, dispatch rate can go up dramatically if technicians are not dispatched with the appropriate spares to clear the problems. Dispatch rate strongly affects staffing required for maintenance and is a major consideration in maintenance costs.

Summary

We have reviewed the basic principles of reliability. It is related to system component failure rates. The way it is measured is dependent on whether the user is interested in how the system serves the body of users or how it serves individual users. Availability depends on failure rate and repair time.

The reliability of the system can be no better than the hardware used in the system. Except in rare cases where little quality control is imposed, software bugs that reach the field generally do not cause major outages. They most often cause unanticipated results and slow system operation. The hardware failures that occur are caused either by conditions that shorten the life of components, or the components reaching the end of their useful life.

A significant contributor to low reliability is low-quality power being supplied to the system. Power failures can contribute to system outages and single-user failure rates. In addition, "dirty" power can shorten the life of system hardware and system components. Designers can create designs that accommodate minor power excursions but wide variations will affect downstream operations.

Similarly, the physical environment in which the equipment operates can shorten the life of components. High humidity and heat are the main culprits, but exposing the system to extreme cold followed by power-up can severely stress the components. PBX vendors specify the normal operating environmental limits for their systems. They also specify short-term operating limits that have a significantly wider range. Operations under the short-term environmental limits for longer than specified will markedly shorten the component life.

The use of redundancy to improve reliability often is misunderstood. Redundancy adds more components to the system and increases system complexity. The overall failure (the MTBF) theoretically, and practically, will increase. What redundancy does is to reduce the repair time (the MTTR). Service is restored quickly, perhaps in zero time. In a well-designed system, the intent of system redundancy is to reduce the MTTR dramatically while modestly increasing the MTBF. Thus, the availability is significantly improved. It should be recognized that strategically placed redundant components can significantly decrease the outage rate (the MTBO).

Security and Fraud Control

Toll fraud has been a longstanding problem for the telecommunications industry. The long distance network has been a target for people seeking financial gain or just beating the system for several years. The earliest broad-based attempts to obtain long distance service fraudulently were probably the "Blue Boxes" of the early 1970s. "Coin simulators," devices that generated false tones indicating that coins were being deposited in pay telephones, were also used during this period. Whereas blue boxes were used sometimes to make hundreds of calls from a single location, coin simulators were used to generate only a few calls from any pay telephone. The absence of a large number of coins from any single location was easily detected and the phone could be watched to catch the perpetrators.

The motivations for toll fraud have been the same throughout its history: money and ego. The money motivation is, first, to receive "free" service for personal use or to resell it. Secondly, it also can be an attempt to hide larger and more sinister activities like drug trafficking and gambling. A high percentage of toll fraud originates in New York City and involves calls to Colombia, the Dominican Republic, and Pakistan. Though the calls originate in New York City, the victims of the fraud can be anywhere in the United States. Access to the distant targeted facilities is gained by using the long distance network.

As the magnitude of the toll fraud problem grew, the IXCs developed strategies to make it more difficult and costly for the perpetrators. The first of these was aggressively investigating cases to catch the perpetrators and then aggressively prosecuting them with as much publicity as possible to discourage others. The second was to "harden" the interexchange system against penetration.

The initial design of the direct-dialed long distance network used tones sent over the voice connection for supervisory signaling. A fairly complete description of these tones and how they were used in the network was published in the November 1960 issue of the *Bell System Technical Journal*. It was virtually a design specification for supervisory toll signal generation. In 1961, the first Blue Box was found at a photography studio at Washington State College.

A 1971 article in a national magazine helped bring Blue Boxes to the public eye. This long and technically accurate article explained how intruders take control of AT&T's network and explained the motivations, penalties, and rewards associated with such activity. The solution to this problem was simple but expensive due to the size of the network. It was Common Channel Interoffice Signaling (CCIS). CCIS removed the supervisory signaling path from the voice path; therefore, the intruders no longer had access to it! It is the equivalent of putting your valuables under lock and key rather than leaving them out in the yard.

Though this largely "solved" the interexchange carrier's problem, it was not because the system became impenetrable! It just became more difficult to attack the network directly than through other more vulnerable endpoints. The first of these was telephone credit cards. Perpetrators would obtain a number from a stolen or compromised credit card (discover the credit card number) and give the number to the operator who had no way to verify it. The call would go through and the perpetrator would disappear. I heard of one person who was given a "random credit card number generator" for a gag going-away gift at Bell Labs!

Another scheme widely used by trucking companies was code calling. A trucker would place a collect call to a number for "Jim Higgens." The call would be refused because there was nobody by that name. The refusing party knew the location the call was placed from and that "Jim Higgens" meant a certain truck was headed to New York with a full load. College students used similar codes to alert friends that they were well or would visit them soon.

Third-party-billed calls (calls placed from one number to a second and billed to a third number) were also a target for abuse. This service is a convenience for callers on the move—to place calls and bill them to their home or business telephone. Abusers would give a third-party number for billing that was not theirs. The discrepancy would not appear until the unsuspecting third party got their bill. For a time, telephone companies would not accept third-party-billed calls unless someone at the third number would verify that the call was legitimate.

Cellular telephone fraud was first noticed about 1986. Carriers found that nonsubscribers were making calls at no charge to themselves by taking advantage of weaknesses in the roamer services. Roamer services permit subscribers to make calls while outside their local serving area. Cellular telephone identification relies on an electronic serial number (ESN) encoded into each cellular telephone. Phones became available on the black market that can change their ESN. These were popular with drug dealers, bookmakers, and other criminals who do not want their activities traceable. Cellular phones are especially effective in this application since they are mobile and very difficult to locate.

The target of telephone fraud most important to this book is PBXs. There are two types of PBX fraud: that where the perpetrator is internal to the location of the PBX and that where the perpetrator is external to the PBX location. Both will be covered, but in the context of the above discussion, the external will be discussed first.

It must be emphasized that the reason PBXs became targets through which to attack the long distance network was that they were more vulnerable targets than others. Not all PBXs are easy targets, just those that are carelessly managed! PBXs are targets of external fraud because some features make it possible to place long distance calls from them without being located on the served premises. These features are Direct Inward System Access (DISA), Voice Mail, and Trunk-to-Trunk Connections. The vulnerability of these features is enhanced by Remote System Maintenance and Administration.

DISA is a feature designed to provide the convenience and economies of the PBX trunking and network to selected individuals when they are not on the served premises. Typically, there is some small number of ports on the PBX designated for remote access. The number ought to be engineered to carry the expected busy hour traffic load at some specified blocking level. Limiting the number of remote access ports serves to throttle the rate at which fraudulent calls can be placed if the access number and passwords are compromised.

When a person with remote access privileges wants to place a long distance call from a remote location, they dial a number that connects them with the remote access port group. Having gained access to a port, they must show that they have remote access privileges by dialing a *Barrier Code*. The maximum length for a barrier code is established by the PBX manufacturer in the design. Following access to the PBX, the user must establish their identity and thus their level of access privileges. This is done by dialing an *authorization code*. Maximum authorization code lengths are also set by the PBX manufacturer. Once this second key has been successfully turned, the user has access to limited PBX privileges as determined by the authorization code dialed. This may be unrestricted access to all PBX services including long distance calling.

Trunk-to-Trunk Connections permits a caller coming in on a trunk in a trunk group to be connected to a trunk in another trunk group. If the second trunk group is a central office or foreign exchange trunk group, the whole world is open! If it is a tie trunk group, the tie trunk may lead to another PBX that allows Trunk-to-Trunk Connections from the tie trunk group to CO trunks. Allowing tie trunks to connect to CO trunks seems safe, since one expects network access to be limited to telephones located on company premises. Therefore, if the call is coming in on a tie trunk, it is assumed to be from a company employee with network privileges. In fact, allowing trunk-to-trunk connections from tie trunks to central office trunks makes the PBX vulnerable to administrative errors at other locations and makes the call source harder to trace.

A trunk-to-trunk connection may be made by dialing the access code for the desired trunk group (most businesses use "9" to access the central office trunks), Call Forwarding from an internal extension, or Call Forwarding from the Voice Mail System (more about that later). Trunk-to-trunk connections also can be set up by an attendant, but for now the attendant is assumed to be trustworthy and knowledgeable. More about this later. A typical scenario for fraud using Trunk-to-Trunk Connections is for a caller to dial into the PBX on a WATS trunk, receive dial tone, and dial "9" plus their number. The victimized PBX owner not only pays for the fraudulent call but for the access call on the WATS trunk!

Voice Mail systems are used as answering machines when the called party does not answer or is busy. Many such systems allow a caller to get dial tone by dialing a simple code that is often included in the greeting message. This facilitates calling someone else after leaving a message or talking to a real person instead of a machine. If restrictions are not placed on what numbers can be called when dial tone is provided, dialing "9" again provides access to the whole world.

Control of access to features and services is done through administration. Barrier Codes, Authorization Codes, Code Restrictions, whether Trunk-to-Trunk Connections are permitted, and most access privileges are controlled through Facility Management. This is the reason companies have more restricted access to Facility Management than Terminal Management. It is easier to open the system gates to intruders from Facility Management. For example, the Facility Restriction Level (FRL) required to access a trunk group or bypass Code Restrictions (Code Restrictions limit which area codes and office codes may be called) could be lowered. Code Restrictions could be changed or eliminated. DISA could be allowed in a system that had elected not to provide this feature. Through Facilities Management, the entire PBX security plan can be compromised.

Because of the power of system administration to change or even eliminate protection from intrusion and fraud, control of system administration access is essential to maintain system security. One of the most vulnerable gateways to the system is the remote maintenance and administration ports. This is typically a dial-up port with password protection. Passwords are sometimes used to define the level of access privileges a particular user has, such as Terminal Management privileges, but not Facility Management privileges.

As mentioned earlier, most PBXs have provisions built in for limiting access and monitoring and logging access activity. Many users will not bother to use these facilities properly and these systems become easy targets for fraud.

Some instances of PBX fraud have resulted in losses of close to one million dollars before they were detected! Many have been more than one hundred thousand dollars. IXCs hold Customer Premises Equipment (CPE) owners responsible for fraud losses originating with their equipment, and the FCC has upheld this position. Several cases are pending in the courts where IXCs have sued CPE owners for fraud losses and where CPE owners have sued IXCs for not alerting them and sued equipment vendors for not providing technical guidance on techniques for fraud control with their equipment. These cases are not settled, but the prospect of such losses should be of concern to most businesses. It is not a realistic goal to prevent all losses but it is realistic to expect to keep these losses small and manageable. One important goal in any security and fraud control plan is to make your system harder to compromise than most. Water runs downhill and crooks run to easy targets!

Fraud Prevention Overview

It is not practical to prevent all fraud within any reasonable economic limits. National defense is really just fraud control against a perpetrator with huge economic resources to devote to the task. No matter how many precautions a company takes, it is possible to defeat the precautions, given enough resources. The economic objective of fraud control is to prevent successful fraud that would cost the perpetrator less than the value of the protected resource.

Preventing fraud must start with a security methodology rather than fraud control techniques. The methodology recommended is:

- Identify the protected resources.
- Write a statement of the threat.
- Define protection methods.
- Define intrusion response.

Let's examine each of these steps.

Identify protected resources

The first step is to identify the resources to be protected. These may include long distance facilities; the administration database; computer ports; and parts of the voice mail system such as mailboxes, its administration database, and call transfer access.

For each protected resource, a statement of policy for the appropriate uses of that resource and who is allowed to use it should be formulated. For example, is it permissible for employees to make local personal calls? What about personal long distance calls?

Each resource should have a statement regarding what level of protection is appropriate. One reasonable but conservative procedure is to access the value of the loss of the protected resource and assume that an intruder would be willing to expend that much to compromise it. For example, if long distance service is compromised, how much service could be stolen before the intrusion is detected? (The answer to this question is not static since the potential loss depends on detection time and time-of-day rate changes.)

Other companies may choose to take a less conservative approach and reduce the level of protection, based on a conservative estimate of the likelihood of penetration. The principal problem with this approach is that the "conservative estimate" cannot be based on much data. It is not a static number, varying with what the rest of the industry is doing for fraud protection. Nevertheless, the balance between the cost of protection and the risk of loss is a management decision.

Write a statement of threat

It is extremely important to write a statement of threat to define what types of intrusions you will defend against. This statement both broadens the scope of intrusions to be considered and narrows the scope of what types of attacks will be defended. Both are important. The statement of threat also helps to clarify the intrusion possibilities and understanding of intrusion techniques. Stating how an intruder will try to compromise the PBX will help to understand any subtleties of the intrusion technique or the system design that make the threat possible. This can lead to simple and effective prevention techniques.

Broadening the scope of intrusions to be defended helps to consider possible sources of attack that may be ignored if the focus is just on defensive measures. For example, does the threat include people internal to the PBX serving location making personal long distance calls? Are the physical premises secure from intrusion by unauthorized people? Are the physical wires between the PBX and the wire center secure? Who are or must be trusted personnel? How trusted are they? Answers to these and similar questions begin to determine what kinds of techniques are needed to prevent or limit losses from these intrusions.

Narrowing the scope of intrusion types defended allows protection resources to be focused on those threats considered most likely or potentially most costly. This is a judgment that is not static. One principal source of information regarding which threats are more likely than others is experience, both yours and other companies'. PBX user groups are an excellent source of information on threats to any specific type of PBX. They are also an excellent source of information on new and emerging threats. There are many clever people out there trying to discover ways to compromise communication resources.

When a new threat is discovered, each company should consider their vulnerability to it and take appropriate countermeasures. These begin with modifying the statement of threat, if necessary, and then developing and implementing effective defenses.

Narrowing the scope of the threat also serves to inform everyone concerned regarding the limits of threat protection. It provides a forum for identifying additional threats, if appropriate, and accessing the system vulnerability to any new threats identified in the telecommunications community. This way, if a company is an early victim of a new threat, everyone understands that this was a limitation of the existing protection techniques and that the security plan may need updating.

Define protection methods

For each element in the threat definition, protection methods must be defined. The protection methods selected should be consistent with the value of the resource at risk. For example, the protected resource may be the administration database, and the threat may be a hacker with automated repetitive dialing capability, a tone/disconnect test to determine the length of the password, and a "random" automated search for the password.

In this scheme, a hacker "discovers" the telephone number for the system's remote maintenance and administration port. Upon dialing, they wait for a second tone or audible signal to signify that the system is ready to accept the dialing of the password. When this tone is received, digits are entered, with a short pause between them, to allow the system to respond when the incorrect password is dialed. When this password rejection response is detected, the number of dialed digits is counted to determine the length of password that the system is expecting.

Subsequently, not necessarily on the same day (or night), the caller calls the port again and begins trying passwords. The password trial scheme is not completely random, in that more-likely strings are attempted first. For example, if the vendor's default password is known, it would be first because many PBX owners will not bother to change it. Next are probably combinations of the numbers in the remote maintenance and administration port telephone number, followed by the Listed Directory Number and the vendor maintenance number if it is known. The point is that *the number of likely choices of passwords is much less than the number of possible choices,* and a smart intruder will try these numbers first. People tend to choose passwords that are easy to remember.

Protection methods against this threat are suggested immediately by the threat statement itself. The first and most obvious is to use password protection and select truly random numbers or characters for the password. A password of at least six digits makes breaking it by Monte Carlo techniques very difficult. (Random searches usually use computers to make them faster and less boring.)

The number doesn't have to be completely random but should have a very nonobvious interpretation to the outsider. Totally random passwords create their own security problems. They are hard to remember, so they are written down. This creates a requirement for physical security, and this leads to inconvenience that promotes carelessness.

Passwords also should be changed periodically to reduce the time window for discovery. The longer a password is used, the longer a perpetrator has to discover it. Discovery takes time. When it is discovered, the perpetrator has more time to exploit it if it is not changed frequently. Both considerations mean that the likelihood of loss increases as the time between password changes increases.

To make passwords effective, it is necessary to have a good Key Management System in place. Key Management is the generation, distribution, and accounting for system security keys (passwords) for the system. There are always the arguments regarding centralized control versus distributed control. Specifically, should the keys be generated by a central authority or should they be selected by the individuals who will use and be responsible for them?

My own preference is for a central facility because it is easier to assure that nontrivial keys are selected. If one key is compromised, everything that the key opens is compromised. A central facility also makes it easier to assure that keys are periodically changed, limiting the time window for discovery. As a key ages, the probability that it has been compromised increases. This is due to several reasons. Hackers have longer to discover the key. People leave and no longer need the key but may use it inappropriately. All other things being equal, the probability of loss of physical security (someone writing it down and losing it or otherwise allowing a perpetrator to see it) increases linearly with time. It is especially important to change a password when someone who knows it no longer needs it.

Key management should be user-friendly. It is one thing to have a secure system defined. It is another to have a system people will use.

The system should be easy to use for its customers, the people who have to use the keys. It should quickly provide keys to those who have a legitimate need for them. Key distribution must be secure and records should be maintained of everyone who has knowledge of a key. These records become the database for determining when a key should be changed because of changing personnel assignments.

The system should *not* provide a prompt, such as a tone, when it is ready to receive a password. Tones or audible prompts just make it easy for people or hardware to determine that a system access code has been dialed. This is especially useful for automated searches through long lists of telephone numbers.

Next, the system should disconnect a caller where two successive passwords are incorrect; the event of such failure should be logged as an intrusion attempt. Two failures is a reasonable choice since it allows for human error in dialing, but not repetitive errors. When a person realizes they have made a dialing error, hopefully they will be more careful on the second try. If Automatic Number Identification (ANI) or ISDN is available, it can be used on the call to identify the source of the intrusion attempt. These intrusion event logs should be examined regularly to discover if the system is under attack or has been successfully compromised.

How often should they be reviewed? At a minimum, an interval such that, if the system was compromised just after the last inspection, you could afford the cost of the maximum loss until the next inspection. Economically, when the potential loss since the last inspection times a conservative estimate of the probability of it happening is equal to the cost of inspecting the log. The probability of such intrusion is probably not known; therefore some would say to assume the worst, a probability of one. If the inspection is manual, this may lead to the impractical solution of having the logs reviewed continuously.

If this is the reasonable assumption, it suggests having the system provide an automatic alarm whenever an intrusion event is detected. Having the system provide an automatic alarm when intrusion attempts exceed a predetermined threshold is one way of getting virtually continuous monitoring (assuming someone watches the alarm). It is important to use a thresholding scheme and set the threshold through experience to prevent the alarm being tripped often through dialing errors from appropriate use. This just teaches people monitoring the alarm to ignore it. If ANI or ISDN is available on the access line, it can be used to screen where calls originate to determine if they are coming from an expected location.

When a system drops a call after several incorrect password attempts, the caller will sometimes call back and get a few more tries before being dropped again. This can be limited by blocking the port after, say, three such attempts where the password key fails. In such cases where the port is blocked, it should remain blocked for a significant period, like several hours. Manually unblocking the port at the PBX site also should be possible. Any instances where the port is blocked due to password failures should be logged as a threat incident.

Another protection against this threat is to open a remote maintenance and administration port access window only when access by appropriate personnel is expected. Any calls to the access number occurring outside this expected call time window are not answered (a technique I use on my FAX machine all the time). This substantially reduces the vulnerability to this threat. It creates the inconvenience of someone at the PBX site opening and closing the window or the window being operated by an automatic timer. This solution may require human intervention and coordination for access, but allows the technical specialist to be remotely located and still perform the maintenance and administration functions.

While these preventive measures are effective against the stated threat, they are ineffective against the threat of a disgruntled or dishonest insider who wishes to compromise the system. This person has knowledge of the access window constraints, the password length, and maybe the password itself. The one possibly effective measure mentioned against this threat is the use of ANI or ISDN for call location screening. Other defenses against this particular threat are simple to devise. This illustrates the importance of identifying all threats that the resource is to be protected against. The above intrusion-protection practices were effective against the outsider, but not the insider.

Define the intrusion response

For each element in the Statement of Threat, protection mechanisms are defined. For each protection mechanism, there needs to be a means of determining when it is being invoked and gathering data about the nature of the intrusion (e.g., reason for invocation and time). This data may be useful in invoking additional protection measures, identifying the perpetrator, or determining that the apparent intrusion attempt was human errors made during appropriate use of the system.

Additional protection of a remote maintenance and administration port might be provided by only allowing access when remote operations are actually expected if a hacker is detected. A perpetrator might be identified by tracing ISDN records of the calling party number. Or, the access attempt may just have been someone making successive dialing errors when beginning a legitimate operation.

Whatever the source of the intrusion attempt, you must know that something is happening in order to do anything about it. If the system does not provide data and alarms when intrusion attempts occur, the

IXCs will, when the intrusion is successful, in the form of a big, long bill! It should always be remembered that any system can be broken. The objectives of security measures are to make that very difficult and to limit losses when it does occur.

Action to be taken in the event of any particular attempted intrusion should be decided ahead of time. This allows for careful planning for effective response while not in the heat of a crisis. The response statement should include any additional data-gathering measures deemed appropriate and any people or organizations that should be notified, including any information that needs to be supplied with the notification.

Operational activities should include a regular review of long distance bills and traffic data on trunks to detect any unexpected changes in calling patterns. Any such change should be immediately investigated to determine if it is legitimate. If the change is sufficiently suspicious, steps should be taken before the investigation to prevent further loss. One company discovered that their PBX had been compromised when a technician with their IXC noticed many calls to Colombia over a weekend and called to ask if these were legitimate. Loss was stopped by not allowing any toll calls until other less drastic measures could be put in place.

Other threats

The hacker breaking into the inadequately protected PBX is the threat that has received the most attention in the past few years. It is not the only serious threat faced by PBX owners. Others include:

- an insider who uses services inappropriately
- a person who gains physical access to transmission circuits
- a social intruder who gains trust through deception

The threat from insiders comes in many forms. With any security system, there is the concept of "trusted individuals." There must be trusted individuals because the system is operated by people for other people. *The degree of trust is not the same for all because it does not need to be.* Trust should not be granted because of "trustworthiness," but because of what is required to do the job. One job qualification is that the person must be worthy of the trust that is required for the job. The administrator who performs Facility Management is more trusted than the most trusted station user (regarding telecommunications resources), not because they are more trustworthy, but because they have to be to do their job. What this means is that if they violate their trust, more assets are at risk. They may or may not deserve the level of trust they are granted. A maintenance technician or attendant may have much trust and may sell access to the system services to get money for a nice Christmas for his or her family.

Protection against loss from trusted insiders begins with adequate screening of those to be entrusted. Past criminal, credit, and employment histories should be obtained and verified. Evaluation of these data is a management responsibility. The evaluation ultimately is a judgment issue involving the history data, character judgment, and legal considerations.

Initial screening is not sufficient. Times change, people change, and opportunities change. Therefore, periodic auditing of system usage patterns and individual use patterns help to identify when inappropriate use begins. Audits should be designed to monitor the use of all levels of privilege and trust. This includes both station users and system administration and operations staff.

Many employees develop the view that free toll calls are a fringe benefit. They may be, but usually are not. This is probably the single area of greatest loss from telecommunications fraud from internal users. The view is reinforced when people use the "benefit" and no one complains. One effective technique of monitoring individual station users uses the Message Detail Recording reports (see Chap. 5, "System Administration, Maintenance, and Reliability"). These reports list all the toll calls for each individual user. They should be reviewed by each person's manager to verify that all the calls are consistent with the stated appropriate-use guidelines. Early feedback to such individuals can usually stop such inappropriate use without undue embarrassment.

Use of privileges by system administration and operations staff should be monitored from several perspectives. The first of these is the data in the database itself. Privileges assigned, FRLs required for use of trunk groups and routing patterns should be verified, Code Restrictions should be verified, and new users should be checked to verify that they are legitimate and that the privileges are properly administered. An easy way to monitor these without being overwhelmed with data is on an exception basis. Once the initial database is verified, only the changes need be reviewed. When using an exception technique, the database from which exceptions are identified and the means of testing against it must have good integrity. This includes accuracy and protection from tampering.

Another perspective is the actual use of the protected resource. Significant deviation from typical usage and usage patterns of protected resources is often the first source of information that an intrusion has occurred. This is an area where there is a crossover between intrusion detection for intrusions from internal and external perpetrators. I have been amazed at how consistent traffic data is from day to day and week to week for any given PBX! The size of the variations is dependent on the particular enterprise served, but it is typically only 5 percent to 10 percent. A location's traffic patterns can be established through measurement and then used as a basis for generating exceptions when a significant deviation is observed. A deviation is not necessarily an intrusion, but a noteworthy incident that should be explained.

Another perspective is frequency and time of use of passwords and other access authenticators. Audit trails of use that includes the access time can provide an alert when these are unusual. These are especially useful for detecting if a password has been compromised.

One of the most important activities that must take place as part of intrusion response is identifying new threats. Whether from an external or internal source, when a new threat is identified, it must be accessed. When appropriate, it should be added to the Statement of Threat, and effective countermeasures developed and put into place. Appropriate short-term countermeasures may be more drastic and inconvenient to the user community than long-term countermeasures. For example, all toll calls may require attendant completion before more subtle countermeasures are developed. The tradeoff between convenience and loss protection is a management decision.

Summary

Telecommunications fraud is a problem that has been around for several decades. The nature of the fraud has evolved with the changing countermeasures and the evolution of the telecommunications system itself. The dollar magnitude of fraud has optimistically not changed over this period and pessimistically has dramatically increased. Only the specific targets and the intrusion techniques have changed.

The motivations for fraud are basically the same: money and ego. Perpetrators obtain financial gain through avoiding or reducing communications bills and through making other illegal activities, such as drug dealing and bookmaking, harder to trace. The ego rewards are largely the satisfaction of beating the system.

As techniques for detecting fraud and identifying perpetrators have improved in some parts of the telecommunications system, perpetrators have moved to more vulnerable targets. One of these is the PBX. PBXs are vulnerable because many owners, either through ignorance or carelessness, do not use the protection facilities provided. This can result in losses of several hundred thousand dollars before a penetration is discovered.

It is not economically feasible to protect a PBX against all possible intrusions. The goal of system security should be to make penetration difficult and, when it does occur, to limit losses to a small manageable amount. This can be accomplished through developing and implementing a comprehensive security plan.

The main elements of the security plan are:

- Identify the protected resources.
- Write a statement of the threat.
- Define protection methods.
- Define intrusion response.

This plan provides a forum for collectively defining what are appropriate protection strategies and quickly identifying system vulnerability to any new threats that develop as the perpetrator community develops new intrusion techniques. Implementation of a sound security plan is effective because it makes intrusion expensive and difficult, therefore not cost-effective. It is especially not cost-effective for the perpetrators because many other businesses will not take such precautions, making them easier targets.

Index

ABOUT THE AUTHORS

STANLEY E. BUSH is widely recognized as an expert in PBX systems and has had a distinguished career at Bell Labs. He is currently director of the Telecommunications Systems Laboratory at the University of Colorado.

CHARLES R. PARSONS is recently retired from Bell Laboratories after 28 years. He currently operates a telecommunications consulting firm specializing in training courses for client staff, and consulting in the areas of PBX system configuration, system administration, and traffic engineering.